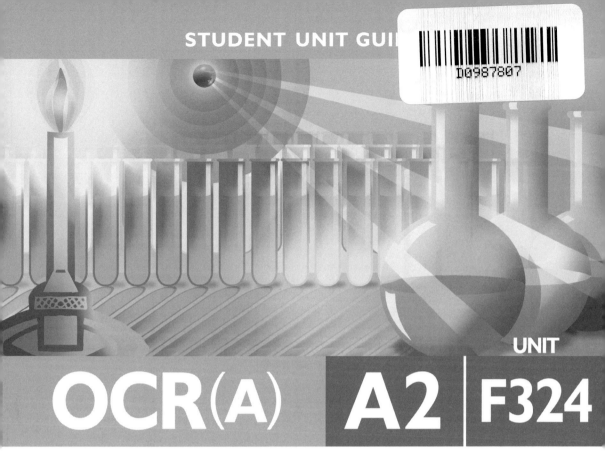

UNIT

OCR(A) A2 F324

Chemistry

Rings, Polymers and Analysis

Mike Smith

Philip Allan Updates, an imprint of Hodder Education, an Hachette UK company, Market Place, Deddington, Oxfordshire OX15 0SE

Orders

Bookpoint Ltd, 130 Milton Park, Abingdon, Oxfordshire OX14 4SB
tel: 01235 827720
fax: 01235 400454
e-mail: uk.orders@bookpoint.co.uk

Lines are open 9.00 a.m.–5.00 p.m., Monday to Saturday, with a 24-hour message answering service. You can also order through the Philip Allan Updates website: www.philipallan.co.uk

ISBN 978-0-340-95758-5

First printed 2009
Impression number 5 4 3 2 1
Year 2014 2013 2012 2011 2010 2009

This guide has been written specifically to support students preparing for the OCR Specification A A2 Chemistry Unit F324 examination. The content has been neither approved nor endorsed by OCR and remains the sole responsibility of the author.

Typeset by Tech-Set Ltd, Gateshead, Tyne and Wear
Printed by MPG Books, Bodmin

Hachette UK's policy is to use papers that are natural, renewable and recyclable products and made from wood grown in sustainable forests. The logging and manufacturing processes are expected to conform to the environmental regulations of the country of origin.

Contents

Introduction

■ ■ ■

Content Guidance

■ ■ ■

Questions and Answers

Introduction

About this guide

This unit guide follows on from a series of two, which together cover the whole OCR AS chemistry specification A. This guide is written to help you prepare for the Unit F324 test: **Rings, Polymers and Analysis**.

This **Introduction** provides advice on how to use the guide, together with suggestions for effective revision.

The **Content Guidance** section gives a point-by-point description of all the facts that you need to know and concepts that you need to understand for Unit F324. It aims to provide you with a basis for your revision. However, you must also be prepared to use other sources in your preparation for the examination.

The **Question and Answer** section shows you the sort of questions you can expect in the unit test. It would be impossible to give examples of every kind of question in one book, but the questions used should give you a flavour of what to expect. Each question has been attempted by two candidates, Candidate A and Candidate B. Their answers, along with the examiner's comments, should help you to see what you need to do to score marks — and how you can easily *not* score marks even though you probably understand the chemistry.

What can I assume about the guide?

You can assume that:

- the topics covered in the Content Guidance section relate directly to those in the specification
- the basic facts you need to know are stated clearly
- the major concepts you need to understand are explained
- the questions at the end of the guide are similar in style to those that will appear in the unit test
- the answers supplied are genuine, combining responses commonly written by candidates
- the standard of the marking is broadly equivalent to the standard that will be applied to your answers

What can I *not* assume about the guide?

You must *not* assume that:

- every last detail has been covered
- the way in which the concepts are explained is the *only* way in which they can be presented in an examination (concepts are often presented in an unfamiliar situation)

- the range of question types presented is exhaustive (examiners are always thinking of new ways to test a topic)

So how should I use this guide?

The guide lends itself to a number of uses throughout your course — it is not *just* a revision aid. The Content Guidance is laid out in sections that correspond to those of the specification for Unit F324, so that you can:

- use it to check that your notes cover the material required by the specification
- use it to identify strengths and weaknesses
- use it as a reference for homework and internal tests
- use it during your revision to prepare 'bite-sized' chunks of related material rather than being faced with a file full of notes

The Question and Answer section can be used to:

- identify the terms used by examiners in questions and what they expect of you
- familiarise yourself with the style of questions you can expect
- identify the ways in which marks are lost as well as the ways in which they are gained

Study skills and revision techniques

All students need to develop good study skills and this section provides advice and guidance on how to study A2 chemistry.

Organising your notes

Chemistry students often accumulate a large quantity of notes, so it is useful to keep these in a well-ordered and logical manner. It is necessary to review your notes on a daily or weekly basis, maybe rewriting the notes you take during lessons so that they are clear and concise, with key points highlighted. It is a good idea to check your notes using textbooks and fill in any gaps. Make sure that you go back and ask the teacher if you are unsure about anything, especially if you find conflicting information in your class notes and textbook.

It is essential to file your notes in specification order, using a consistent series of headings. The Content Guidance section can help you with this.

Organising your time

When organising your time, make sure that you plan carefully, allowing enough time to cover all the work. It sounds easy, but it is one of the most difficult things to do. There is considerable evidence to show that revising for 2–3 hours at a time is counter-productive and that it is better to work in short, sharp bursts of between 30 minutes and an hour.

Preparation for examinations is a very personal thing. Different people prepare, equally successfully, in very different ways. The key is being honest about what *works for you.*

Whatever your style, you must have a plan. Sitting down the night before the examination with a file full of notes and a textbook does not constitute a revision plan — it is just desperation — and you must not expect a great deal from it. Whatever your personal style, there are a number of things you *must* do and a number of other things you *could* do.

Unit F324 contains lots of new terminology and concepts. There is also a large body of factual knowledge, relating to reagents and reaction conditions, which must be learnt.

The scheme outlined below is a suggestion as to how you might revise Unit F324 over a 3-week period. The work pattern shown is fairly simple. It involves revision and/or rewriting a topic and then over the next few days going through it repeatedly, but never spending more than 30 minutes at a time. When you are confident that you have covered all areas, start trying to answer the questions from past papers, from the Question and Answer section, or from the A2 OCR textbook. Mark them yourself and seek help with anything that you are not sure about.

Day	Week 1	Week 2	Week 3
Mon	Topic 1: Arenes — allow about 20 minutes	Make summary notes on each of: Topic 1: Arenes Topic 2: Carbonyls Topic 3: Carboxylic Acids, Esters and Amines Topic 4: Amino Acids, Proteins and Chirality Topic 5: Polymers and Synthetic Routes Topic 6: Chromatography and Spectroscopy	Use your past papers or questions from your textbook(s) and try one structured question on Topic 6 Mark it and list anything you don't understand Allow about 30 minutes
Tues	Topic 2: Carbonyls — allow about 20 minutes Reread yesterday's notes on Topic 1 — 10 mins	Reread all your summary notes at least twice	You have now revised all of Unit F324 and have attempted questions relating to each topic Make a list of your weaknesses and ask your teacher for help Reread all your summary notes at least twice Ask someone to test you

Day	Week 1	Week 2	Week 3
Wed	Topic 3: Carboxylic Acids, Esters and Amines Allow about 20 mins Reread yesterday's notes on Topics 2 and 5 — 10 mins Go over Topic 1 again — 5 mins	Use your past papers or questions from your textbook(s) and try one structured question on Topic 1 Mark it and list anything you don't understand Allow about 30 minutes	Use your past papers or textbooks and try relevant questions that require extended writing (essay-type questions) from any one of the seven topics Mark it and list anything you don't understand Allow about 30 minutes
Thurs	Topic 4: Amino Acids, Proteins and Chirality Allow about 20 mins Reread yesterday's notes on Topic 3 — 10 mins Go over Topic 2 — 5 mins Look again at Topic 1 — 2 mins	Use your past papers or questions from your textbook(s) and try one structured question on Topic 2 Mark it and list anything that you don't understand Allow about 30 minutes	Reread all your summary notes at least twice Concentrate on the weaknesses you identified on Monday (by now you should have talked to your teacher about them) Get someone to test you
Fri	Topic 5: Polymers and Synthetic Routes Allow about 20 mins Reread yesterday's notes on Topic 4 — 10 mins Go over Topic 3 — 5 mins Finally, look over Topic 2 — 2 mins By now you should know Topic 1	Use your past papers or questions from your textbook(s) and try one structured question on Topic 3 Mark it and list anything that you don't understand. Allow about 30 minutes	Timed exam practice Allow 1 hour and attempt a past exam paper. Use your notes and textbooks to mark your responses List anything that you don't understand See your teacher for additional help with your weaknesses
Sat	Topic 6: Chromatography and Spectroscopy Allow about 20 mins Reread yesterday's notes on Topic 5 — 10 mins Go over Topic 4 — 5 mins Finally, check over Topic 3 — 2 mins By now you should know Topics 1 and 2	Use your past papers or questions from your textbook(s) and try one structured question on Topic 4 Mark it and list anything you don't understand. Allow about 30 minutes	Timed exam practice. Allow 1 hour and attempt a past exam paper Use your notes and textbooks to mark your responses. List anything that you don't understand See your teacher for additional help with your weaknesses

Day	Week 1	Week 2	Week 3
Sun	Allow about 20 mins in total. Rereading yesterday's notes on Topic 6 — 10 mins Go over Topic 5 — 5 mins Finally, check over Topic 4 — 2 minutes By now you should know Topics 1–3	Use your past papers or questions from your textbook(s) and try one structured question on Topic 5 Mark it and list anything that you don't understand. Allow about 30 minutes	Timed exam practice. Allow 1 hour and attempt a past exam paper Use your notes and textbooks to mark your responses List anything that you don't understand See your teacher for additional help with your weaknesses
Mon	Topic 2: Carbonyls — allow about 20 minutes Reread yesterday's notes on Topic 1 — 10 mins	Reread all your summary notes at least twice	You have now revised all of Unit F324 and have attempted questions relating to each topic Make a list of your weaknesses and ask your teacher for help Reread all your summary notes at least twice Ask someone to test you

The revision timetable may not suit you, in which case write one to meet your needs. It is only there to give you an idea of how one might work. The most important thing is that the grid at least enables you to see what you should be doing and when you should be doing it. Don't try to be too ambitious — *little and often is by far the best way*.

It would of course be much more sensible to put together a longer, rolling programme to cover all of your A2 subjects. Do *not* leave it too late. Start sooner rather than later.

Things you *must* do

- Leave yourself enough time to cover *all* the material.
- Make sure that you actually *have* all the material to hand (use this book as a basis).
- Identify weaknesses early in your preparation so that you have time to do something about them.
- Familiarise yourself with the terminology used in examination questions.

Things you *could* do to help you learn

- Copy selected portions of your notes.
- Write a précis of your notes, which includes all the key points.
- Write key points on postcards (carry them round with you for a quick revise during a coffee break).

- Discuss a topic with a friend also studying the same course.
- Try to explain a topic to someone *not* on the course.
- Practise examination questions on the topic.

Approaching the unit test

Terms used in the unit test

You will be asked precise questions in the examinations, so you can save a lot of valuable time as well as ensuring you score as many marks as possible by knowing what is expected. Terms most commonly used are explained below.

Define

This requires a precise statement.

Explain/what is meant by?

This normally implies that a definition should be given, together with some relevant comment on the significance or context of the term(s) concerned, especially where two or more terms are included in the question. The amount of supplementary comment intended should be determined by the mark allocation.

State

This implies a concise answer with little or no supporting argument.

Describe

This requires candidates to state in words (but using diagrams where appropriate) the main points of the topic. It is often used with reference to mechanisms and requires a step-by-step breakdown of the reaction including, where appropriate, curly arrows to show the movement of electrons. The amount of description intended should be interpreted in the light of the mark allocation.

Deduce or predict

This implies that you are not expected to produce the required answer by recall but by making a logical connection between other pieces of information. Such information may be wholly given in the question or could depend on answers given in an earlier part of the question. 'Predict' also implies a concise answer with no supporting statement required.

Outline

This implies brevity, i.e. restricting the answer to essential detail only.

Suggest

This is used in two contexts. It implies either that there is no unique answer, or that you are expected to apply your general knowledge to a 'novel' situation that may not be formally in the specification.

Calculate

This is used when a numerical answer is required. In general, working should be shown.

Sketch

When applied to diagrams, this means that a simple, freehand drawing is acceptable. Nevertheless, care should be taken over proportions, and important details should be labelled clearly.

On the day

When you finally open the test paper, it can be quite a stressful moment and you need to be certain of your strategy. The test paper will consist of structured questions (usually about four or five) and free-response questions. The structured questions will usually account for between 45 and 55 marks and the free-response section will be worth between 5 and 10 marks. The total number of marks on the paper is 60.

Time will be tight; there are only 60 minutes for this 60-mark paper. So please:

- do *not* begin to write as soon as you open the paper
- scan *all* the questions before you begin to answer any
- identify those questions about which you feel most confident
- *read the question carefully* — if you are asked to explain, then explain, do *not* just describe
- take notice of the mark allocation and don't supply the examiner with all your knowledge of any topic if there is only 1 mark allocated (similarly, you will have to come up with *four* ideas if 4 marks are allocated)
- try to stick to the point in your answer (it is easy to stray into related areas that will not score marks and will use up valuable time)
- try to answer *all* the questions

Structured questions

These are short-answer questions that may require a single-word answer, a short sentence or a response amounting to several sentences. The setter for the paper will have thought carefully about the amount of space required for the answer and the marks allocated, so the space provided usually gives a good indication of the amount of detail required.

Free-response questions

These questions enable you to demonstrate the depth and breadth of your knowledge, as well as your ability to communicate chemical ideas in a concise way. These questions will often include marks for the quality of written communication (QWC). You will be expected to use appropriate scientific terminology and to write in continuous prose, paying particular attention to spelling, punctuation and grammar.

Content
Guidance

This Content Guidance section is a student's guide to Unit F324. The topics covered are:

Module 1: Rings, acids and amines
- Arenes
  - structure of benzene
  - electrophilic substitution reactions
  - phenols
- Carbonyl compounds
  - reactions of carbonyl compounds
  - characteristic tests for carbonyl compounds
- Carboxylic acids and esters
  - properties of carboxylic acids
  - esters, triglycerides, unsaturated and saturated fats
- Amines
  - basicity of amines
  - preparation of amines
  - synthesis of azo dyes

Module 2: Polymers and synthesis
- Amino acids, proteins and chirality
  - amino acids
  - peptide formation and hydrolysis of proteins
  - optical isomerism
- Polyesters and polyamides
  - condensation polymerisation
  - hydrolysis and degradable polymers
- Synthesis
  - synthetic routes
  - chirality in pharmaceutical synthesis

Module 3: Analysis
- Chromatography
  - types of chromatography, GC–MS
- Spectroscopy
  - NMR spectroscopy
  - combined techniques

This section includes all the relevant key facts required by the specification and explains the essential concepts.

Required AS chemistry

Unit F324 aims to build upon the concepts developed during the AS course, particularly those first met in Unit F322: Chains, Energy and Resources. You are expected to use the concepts developed at AS organic chemistry. You are *not* expected to recall factual knowledge such as reagents and conditions for specific reactions.

You should be able to recall that organic chemicals can be grouped together to form homologous series, in which one member of the series differs from the next by CH_2. Each member of a homologous series has a different group of atoms attached to the carbon chain or ring. This group determines the chemistry of the series and is referred to as the functional group. A general formula can be written for each homologous series.

You need to be able to use displayed, molecular, empirical, structural and skeletal formulae. You should be able to name individual chemicals, including structural isomers and *E–Z* (formerly known as *trans-* and *cis-*) isomers

Name	Displayed formula	Molecular formula	Empirical formula	Structural formula	Skeletal formula
But-1-ene	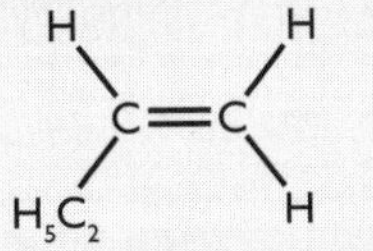	C_4H_8	CH_2	$CH_3CH_2CHCH_2$	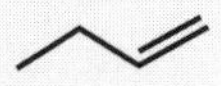
(Z)-but-2-ene (*cis*)	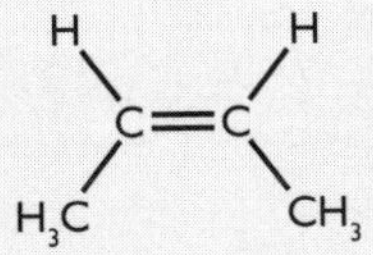	C_4H_8	CH_2	$CH_3CHCHCH_3$	
(E)-but-2-ene (*trans*)	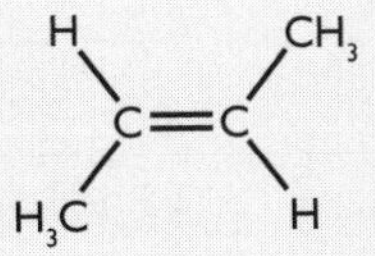	C_4H_8	CH_2	$CH_3CHCHCH_3$	
Methylpropene	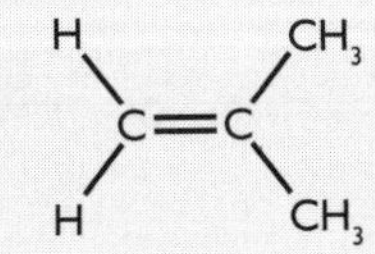	C_4H_8	CH_2	$CH_2C(CH_3)_2$	

You will also be expected to determine empirical formulae and to calculate atom economy and percentage yield from experimental data.

Terms first introduced at AS chemistry will be used and it is essential that these are fully understood. They include the following:
- **radical** — contains a single unpaired electron (e.g. Cl•)
- **nucleophile** — an electron-pair donor (e.g. OH⁻)
- **electrophile** — an electron-pair acceptor (e.g. H⁺)

It is unlikely that you will be tested on the specific mechanisms met in Unit F322. However, the concept of showing movement of electrons by the use of curly arrows is developed further and it is advisable to revise the three mechanisms encountered in AS organic chemistry:
- **radical substitution** in reactions between alkanes and halogens
- **electrophilic addition** in reactions between alkenes and halogens
- **nucleophilic substitution** in the hydrolysis of halogenoalkanes

Addition polymerisation was introduced at AS and is studied in more detail in this A2 unit. It is essential that you fully understand the basic ideas and can construct relevant polymers from a given monomer. Some common monomers and their corresponding polymers are shown below:

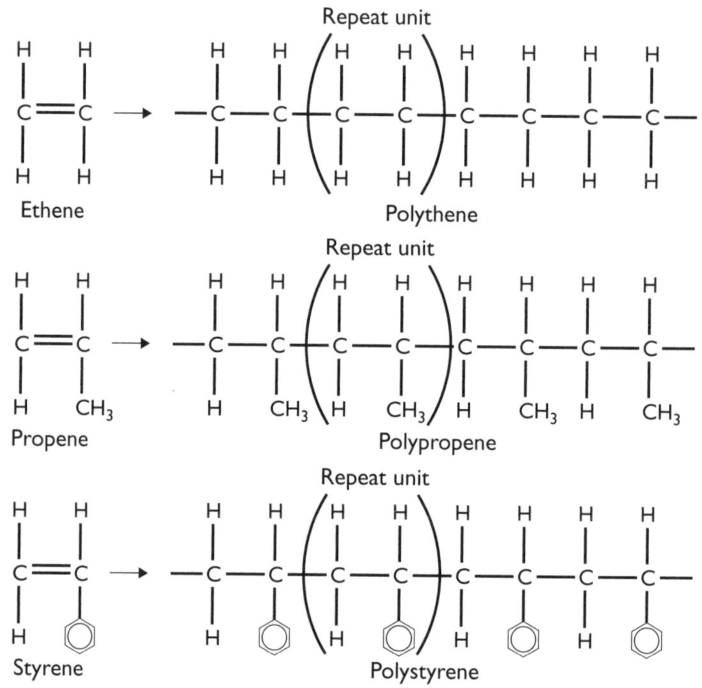

The section on alcohols is particularly relevant, as Unit F324 covers the oxidation of alcohols and looks at all three of the organic products (aldehydes, ketones and carboxylic acids). It is therefore essential that you can recall the relevant equations, reagents and conditions.

Module 1: Rings, acids and amines

Arenes

The simplest arene is benzene. Its composition by mass is C, 92.3% and H, 7.7%. Its relative molecular mass is 78. This information shows that the empirical formula is CH and the molecular formula is C_6H_6.

Structure of benzene

The chemist August Kekulé suggested that benzene was a cyclic molecule with alternating C=C double bonds and C—C single bonds.

However, there are three major pieces of evidence against this type of structure:

- Compounds that contain C=C double bonds readily decolorise bromine. Benzene reacts with bromine only when boiled and exposed to ultraviolet light. This casts doubt on the existence of C=C double bonds in benzene.
- On average, the length of the C—C single bond is 154 pm, while the average length of a C=C double bond is 134 pm. All the bonds in benzene are 139 pm. This suggests an intermediate bond somewhere between a double bond and a single bond.
- Experimentally determined enthalpy changes for cyclohexene and benzene reacting separately with hydrogen give a value for benzene about $150\,kJ\,mol^{-1}$ less than the alternating double-bond–single-bond model.

The current model of benzene takes these observations into account. Each carbon atom contributes one electron to a π-delocalised ring of electrons above and below the plane of atoms. Each carbon has one *p*-orbital at right angles to the plane of atoms.

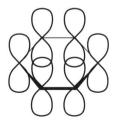

Each *p*-orbital overlaps with adjacent *p*-orbitals, so that delocalisation is extended over all six carbon atoms.

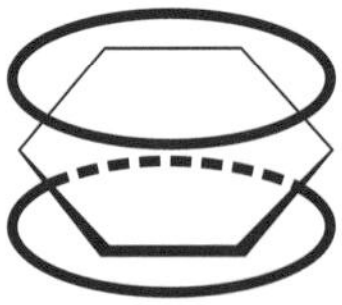

π-delocalised ring above and below the plane

The π-delocalised ring accounts for the increased stability of benzene, as well as explaining the reluctance to react with bromine. In addition, it also explains why all six carbon–carbon bond lengths are identical. Benzene is usually represented by the skeletal formula shown below.

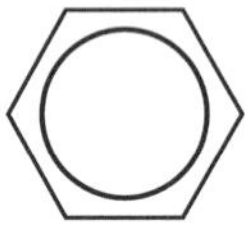

Electrophilic substitution

As a result of the presence of the π-delocalised ring, benzene is very stable. However, benzene does react with electrophiles that have a full positive charge — an induced dipole in a molecule is not normally sufficient.

An electrophile is defined as an electron-pair acceptor. The most common electrophile is H^+, and it is clearly pointless reacting benzene (C_6H_6) with H^+. Catalysts are used to generate electrophiles such as NO_2^+, Cl^+, Br^+ and CH_3^+. The general equation for this is:

$$C_6H_6 + X^+ \rightarrow C_6H_5X + H^+$$

where X^+ is the electrophile.

Nitration of benzene

Reagents: HNO_3 and H_2SO_4 (catalyst)

Conditions: approximately 60°C

Balanced equation: $C_6H_6 + HNO_3 \rightarrow C_6H_5NO_2 + H_2O$

Mechanism:
- Generation of the electrophile:

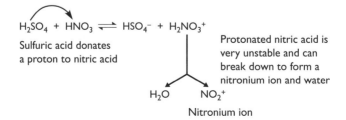

$$H_2SO_4 + HNO_3 \rightleftharpoons HSO_4^- + H_2NO_3^+$$

Sulfuric acid donates a proton to nitric acid

Protonated nitric acid is very unstable and can break down to form a nitronium ion and water

H_2O NO_2^+

Nitronium ion

- Electrophilic attack at the benzene ring:

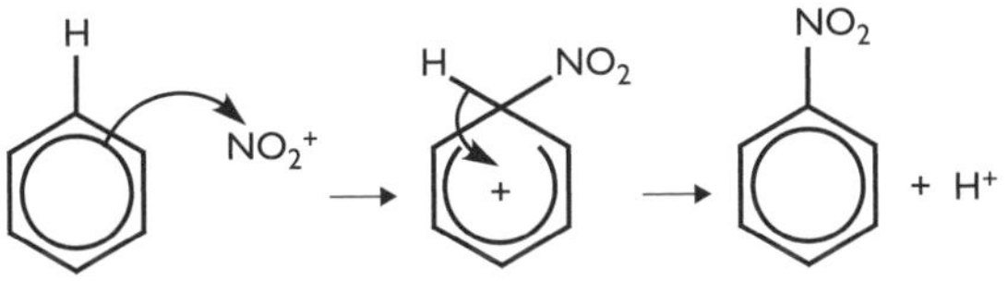

- Regeneration of the catalyst:

$$H^+ + HSO_4^- \rightarrow H_2SO_4$$

Halogenation of benzene

Reagents: Cl_2 and $AlCl_3$ (catalyst)

Conditions: anhydrous

Balanced equation:

$$C_6H_6 + Cl_2 \rightarrow C_6H_5Cl + HCl$$

Mechanism:
- Generation of the electrophile:

$$Cl_2 + AlCl_3 \rightarrow Cl^+ + AlCl_4^-$$

- Electrophilic attack at the benzene ring:

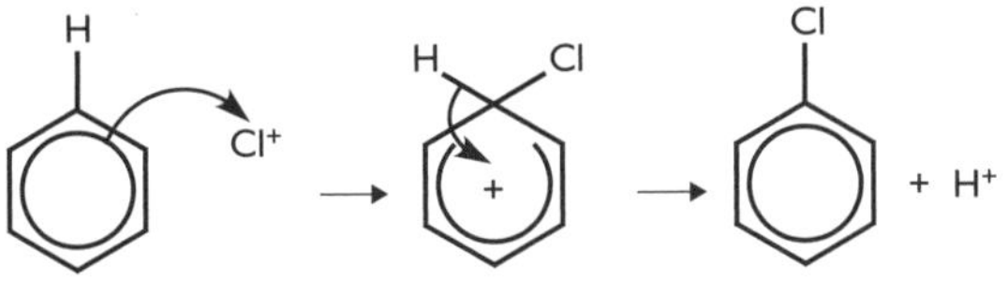

- Regeneration of the catalyst:

$$H^+ + AlCl_4^- \rightarrow AlCl_3 + HCl$$

The chlorination of benzene is referred to as a Friedel–Crafts reaction, in which $AlCl_3$ behaves as a halogen carrier. Halogen carriers are able to accept a halide ion and to 'carry it' through the reaction. At the end of the reaction, the halide ion is released and the hydrogen halide is formed. All aluminium halides, iron(III) halides and iron can behave as halogen carriers.

Bromination of alkenes and arenes

You will recall from Unit F322 that alkenes, such as cyclohexene, react readily with bromine, in the absence of sunlight, and undergo electrophilic addition reactions.

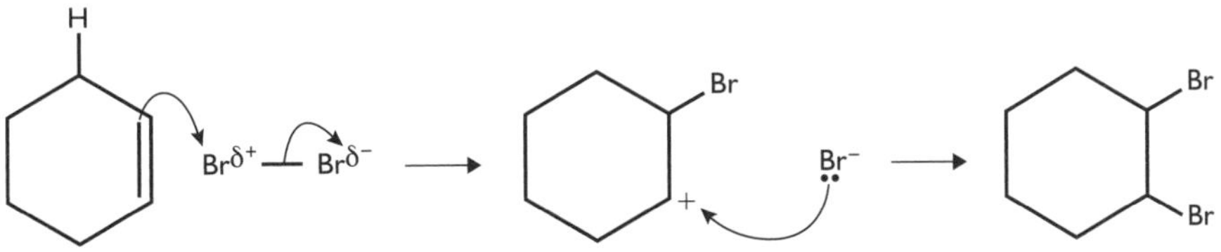

The reaction is rapid and is initiated by the induced dipole in bromine.

Benzene also reacts with bromine but it is more resistant, reacting much less readily than an alkene such as cyclohexene. Benzene requires an electrophile with a full positive charge, Br^+, which is generated in the presence of a halogen carrier. The resultant reaction is electrophilic substitution, *not* electrophilic addition. This is explained by the stability of the π-delocalised ring of electrons that is retained in most reactions of arenes.

The relative ease of reaction of cyclohexene can be explained by the electron density:
- Cyclohexene has a C=C bond, which has a high electron density.
- The electron density in cyclohexene is sufficient to induce a dipole in the Br–Br bond, generating an electrophile.
- The electrophile is attracted to cyclohexene because of the latter's high electron density.

Uses of arenes

Arenes such as benzene, methylbenzene and 1,4-dimethylbenzene are used as additives to improve the performance of petrol. They are manufactured by reforming straight-chain alkanes.

Hexane

C_6H_{14}

$+$ $4H_2$

Octane

C_8H_{18}

$+$ $4H_2$

Benzene is the feedstock for a variety of products, ranging from medicines such as aspirin and benzocaine to explosives such as 2,4,6-trinitromethylbenzene (TNT), and including a wide range of azo dyes. Phenylethene (styrene) is also manufactured from benzene and is the monomer used to produce the polymer poly(phenylethene) or polystyrene.

Products made from benzene are of great value. However, benzene itself is carcinogenic and may cause leukaemia. Chlorinated benzene compounds are extremely toxic.

Phenols

In phenols, the —OH group is attached directly to the benzene ring.

is a phenol but   is an alcohol.

Alcohols, like C_2H_5OH, are soluble in water because they form hydrogen bonds with water. Phenol (C_6H_5OH) is sparingly soluble in water. Although the —OH bond in phenol forms hydrogen bonds with water, the benzene ring reduces the solubility because it forms van der Waals bonds with neighbouring phenol molecules.

Phenols are weak acids, but alcohols are not acidic — ethanol ionises less than water. This difference in acidity is best explained by the relative inductive effect of the aryl and the alkyl groups and the relative stability of the phenoxide and ethoxide ions.

The inductive effect can be regarded as the movement of electrons along a σ-bond. It is caused by differences in electronegativities and electron densities. Alkyl groups, like methyl and ethyl, release electrons along the σ-bond and have a positive inductive effect. In ethanol, this has the effect of increasing the electron density in the O—H bond and hence ethanol is unlikely to donate a H^+ ion. Additionally, the ethoxide ion is made unstable because the positive inductive effect pushes electrons towards the O, making it more likely to accept a proton.

The benzene ring (C_6H_5-) pulls electrons into the ring and can be described as having a negative inductive effect. This weakens the O—H bond in phenol and stabilises the phenoxide ion. It also activates the ring, particularly at the 2, 4 and 6 positions. Consequently, phenol behaves as a weak acid and also undergoes electrophilic substitution reactions much more readily than benzene.

Phenol forms salts by the reaction with both NaOH and with Na:

$$C_6H_5-OH + NaOH \rightarrow C_6H_5-O^- Na^+ + H_2O$$

$$C_6H_5-OH + Na \rightarrow C_6H_5-O^- Na^+ + \tfrac{1}{2}H_2$$

Phenol reacts readily with bromine. The bromine is decolorised and white crystals of 2,4,6-tribromophenol are formed.

Unlike benzene, phenol does not require a halogen carrier and reacts instantly with phenol. This is explained by the activation of the ring due to:
- delocalisation of the lone pairs of electrons on the oxygen atom into the ring
- which increases the electron density, which in turn polarises the halogen
- which increases the attraction for the halogen
- all of which result in an increased reactivity for phenol compared to benzene

Uses of phenols

Phenols are used in the production of disinfectants and antiseptics. Nowadays, chlorophenols such as 2,4,6-trichlorophenol (TCP) are widely used as antiseptics and disinfectants. Phenols are also used in the production of plastics and resins.

Carbonyl compounds

The presence of a carbonyl group (C=O) in a molecule means that it is unsaturated. The position of the C=O on the carbon chain determines whether or not the molecule is classified as an aldehyde or a ketone. Aldehydes always have the C=O at the end of the carbon chain.

Aldehydes   **Ketones**

Propanal   Propanone

Butanal   Butanone

You met the carbonyl group in Unit F322 when you studied the chemistry of alcohols. Aldehydes are formed in the first stages of oxidation of primary alcohols, while ketones are formed when secondary alcohols are oxidised. The oxidising mixture is $Cr_2O_7^{2-}/H^+$ (e.g. $K_2Cr_2O_7/H_2SO_4$), which is bright orange and changes to green during the redox process. When oxidising a primary alcohol, the choice of apparatus is important. Refluxing produces a carboxylic acid, while distillation gives an aldehyde.

Balanced equations can be written for the oxidation reactions by using [O] to represent the oxidising agent.

Oxidation of primary alcohols to aldehydes

$CH_3OH + [O] \rightarrow HCHO + H_2O$
Methanol   Methanal

$CH_3CH_2OH + [O] \rightarrow CH_3CHO + H_2O$
Ethanol   Ethanal

Oxidation of secondary alcohols to ketones

$CH_3CHOHCH_3 + [O] \rightarrow CH_3COCH_3 + H_2O$
Propan-2-ol   Propanone

$CH_3CH_2CH(OH)CH_3 + [O] \rightarrow CH_3CH_2COCH_3 + H_2O$
Butan-2-ol   Butanone

Reactions common to aldehydes and ketones

Reduction

Aldehydes or ketones can be reduced to their respective alcohols. Sodium tetrahydridoborate(III) ($NaBH_4$) is a suitable reducing agent. [H] is used to represent the reducing agent in any balanced equation.

Aldehydes are reduced to **primary alcohols**:

$$CH_3CH_2CHO + 2[H] \rightarrow CH_3CH_2CH_2OH$$

Ketones are reduced to **secondary alcohols**:

$$CH_3COCH_3 + 2[H] \rightarrow CH_3CH(OH)CH_3$$

Nucleophilic addition reactions

The carbonyl group is unsaturated and polar and consequently undergoes nucleophilic addition reactions.

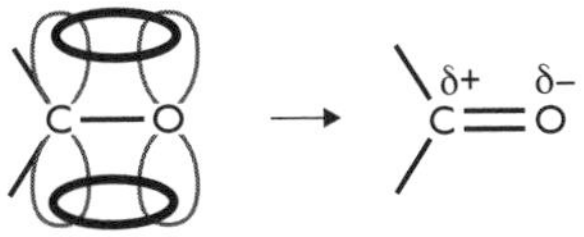

The p-orbitals overlap to form a π-bond

This is best illustrated by the reduction reactions described above. As with all mechanisms, you should show the movement of electrons by using curly arrows. You should also include relevant dipoles and lone pairs of electrons.

$NaBH_4$ is the source of the hydride ion (H^-), which behaves as the nucleophile. The intermediate formed reacts with the solvent (H_2O) to form the alcohol. The alcohol can also be formed by adding an acid, $H^+(aq)$, to the intermediate.

Reduction of ethanal with aqueous NaBH₄

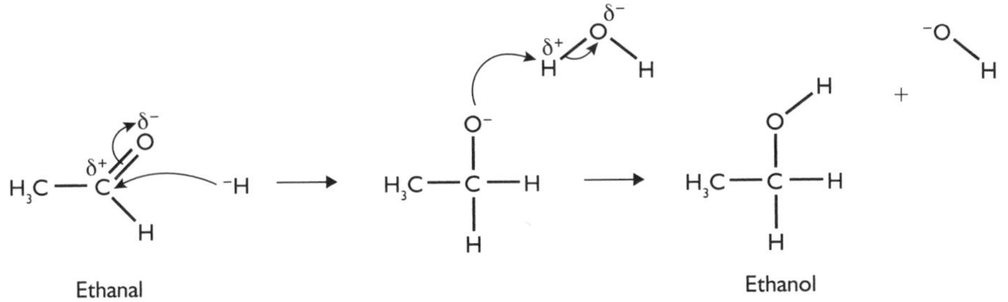

Ethanal   Ethanol

Ketones behave in exactly the same way.

Reduction of propanone with aqueous NaBH₄

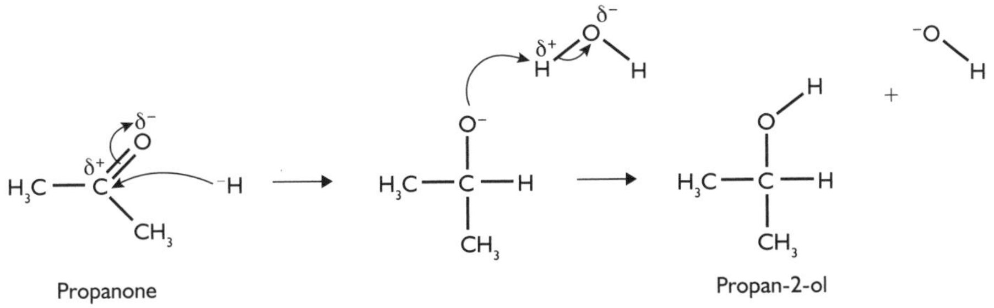

Propanone   Propan-2-ol

Reactions with 2,4-dinitrophenylhydrazine

You are not expected to recall the formula of 2,4-dinitrophenylhydrazine (the abbreviation 2,4-DNPH is often acceptable). Reactions with 2,4-DNPH are important for two reasons:

(1) Carbonyls react with 2,4-DNPH to produce distinctive precipitates. The precipitates are usually bright red, orange or yellow. Therefore, this reaction can be used to identify the presence of a carbonyl group.

(2) The organic product (the 2,4-DNPH derivative) is relatively easy to purify by recrystallisation. Therefore, the melting point of the brightly coloured precipitate can be measured. Each derivative has a different melting point, and the value of the melting point can be used to identify the specific carbonyl. The table below shows the melting points of the derivatives of a few common carbonyl compounds:

Carbonyl compound	Melting point of 2,4-DNPH derivative/°C
Ethanal	142–43
2-Methylpropanal	180–81
Butanone	109–10
3-Methylbutan-2-one	119–20

Reactions of aldehydes alone

Aldehydes and ketones can be distinguished by a series of redox reactions. Aldehydes are readily oxidised to carboxylic acids, whereas ketones are not easily oxidised.

Aldehydes react with Tollens' reagent, which is an aqueous solution of Ag^+ ions in an excess of ammonia, $Ag(NH_3)_2^+$. When Tollens' reagent is reacted with an aldehyde and warmed gently in a water bath at about 60°C, silver metal is precipitated, which forms a distinctive silver mirror. This reaction is a redox reaction, whereby the Ag^+ is reduced to Ag metal, and the aldehyde is oxidised to a carboxylic acid.

The Ag^+ ion gains an electron and is, therefore, reduced to silver

$$Ag^+ + e^- \xrightarrow{\text{Reduction}} Ag \text{ (silver mirror)}$$

The aldehyde is oxidised to a carboxylic acid

Tollens' reagent does not react with a ketone, because ketones are not oxidised.

The oxidation of an aldehyde to a carboxylic acid can also be achieved by refluxing with acidified dichromate ($H^+/Cr_2O_7^{2-}$), where a colour change from orange to green is observed.

The oxidation of an aldehyde can be followed by using infrared spectroscopy, which was first introduced in Unit F322.

Group	Compounds	IR absorption/cm^{-1}
C=O	Aldehydes, ketones, carboxylic acids	1680–1750
O—H	Carboxylic acids	2500–3300

The absorption due to the carbonyl group can clearly be seen in the infrared spectrum of ethanal shown below:

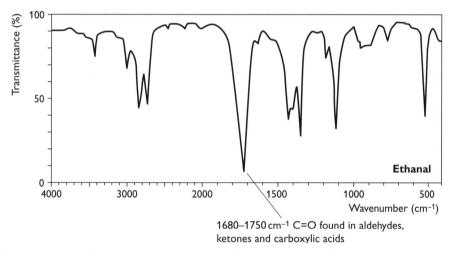

1680–1750 cm^{-1} C=O found in aldehydes, ketones and carboxylic acids

The spectrum below shows the absorption for both C=O and O—H, confirming that the ethanal has been oxidised to ethanoic acid.

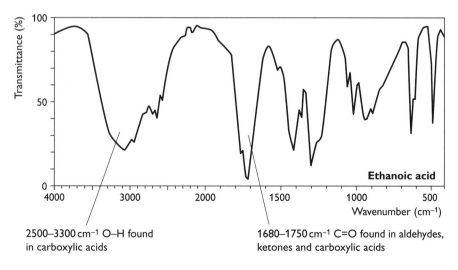

2500–3300 cm^{-1} O—H found in carboxylic acids

1680–1750 cm^{-1} C=O found in aldehydes, ketones and carboxylic acids

Carboxylic acids and esters

All carboxylic acid contain the functional group:

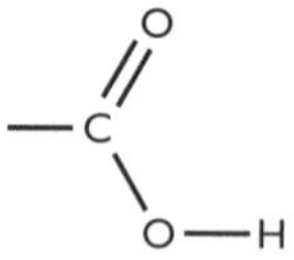

Carboxylic acids such as methanoic acid and ethanoic acid are soluble in water. This solubility can be explained by the ability of carboxylic acids to form hydrogen bonds with water.

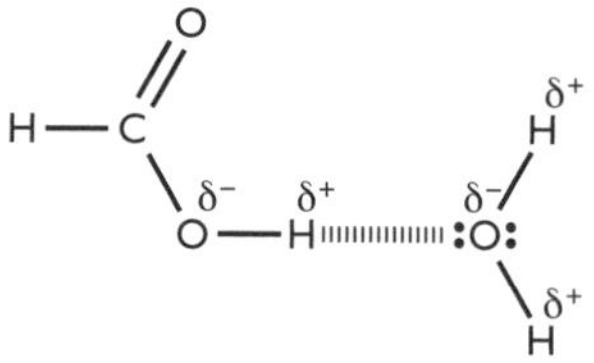

Methanoic acid is soluble in water because it can form H-bonds with water molecules

The solubility of carboxylic acids decreases with increasing molecular mass.

Carboxylic acids are weak acids. They can donate protons, but they only partially dissociate into their ions:

$$CH_3COOH(aq) \rightleftharpoons CH_3COO^-(aq) + H^+(aq)$$

The carboxylic acid group can be attached to either a chain (aliphatic) or to a ring (aromatic).

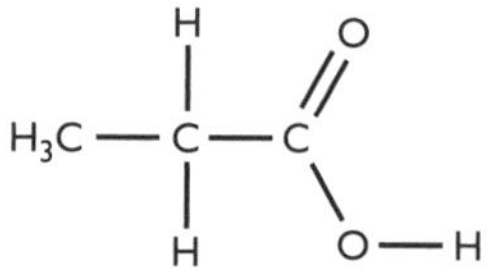

Propanoic acid

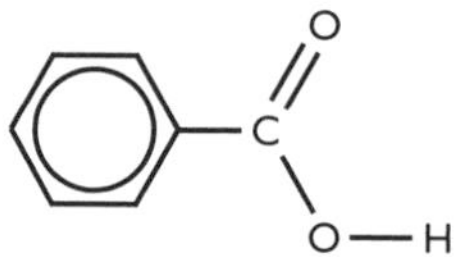

Benzoic acid

Carboxylic acids display typical reactions of an acid and form salts (**carboxylates**). Salt formation can occur by any of the following reactions:

- acid + base → salt + water

$$CH_3COOH(aq) + NaOH(aq) \rightarrow CH_3COO^-Na^+(aq) + H_2O(l)$$

Ethanoic acid   Sodium ethanoate

- acid + (reactive) metal → salt + hydrogen

$$CH_3COOH(aq) + Na(s) \rightarrow CH_3COO^-Na^+ + \tfrac{1}{2}H_2$$

Ethanoic acid   Sodium ethanoate

- acid + carbonate → salt + water + carbon dioxide

$$2CH_3COOH(aq) + Na_2CO_3(aq) \rightarrow 2CH_3COO^-Na^+(aq) + H_2O(l) + CO_2(g)$$

Ethanoic acid   Sodium ethanoate

The reaction with a carbonate can be used as a test for a carboxylic acid. When an acid is added to a solution of a carbonate, bubbles (effervescence) of carbon dioxide are seen.

Carboxylic acids can also react with alcohols to form esters. This type of reaction is known as **esterification**. It is a reversible reaction and is usually carried out in the presence of an acid catalyst such as concentrated sulfuric acid. The general reaction can be summarised as follows:

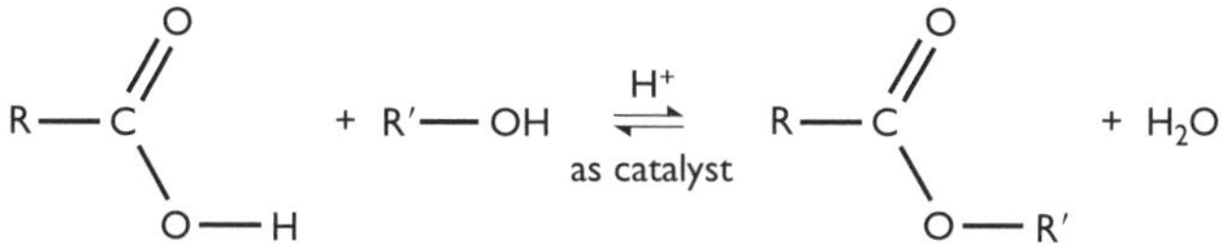

Carboxylic acid   Alcohol   Ester   Water

Esters are named from the alcohol and the carboxylic acid from which they are derived. The first part of the name relates to the alcohol and the second part of the name to the acid. For example:

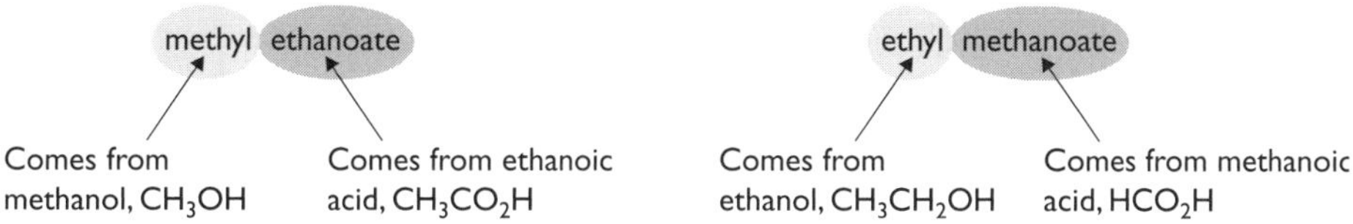

methyl ethanoate		ethyl methanoate	
Comes from methanol, CH_3OH	Comes from ethanoic acid, CH_3CO_2H	Comes from ethanol, CH_3CH_2OH	Comes from methanoic acid, HCO_2H

When organic compounds react, the reaction usually occurs between the two functional groups, in this case the alcohol and the carboxylic acid. It is helpful to draw the two reacting molecules with the functional groups facing each other.

For example:

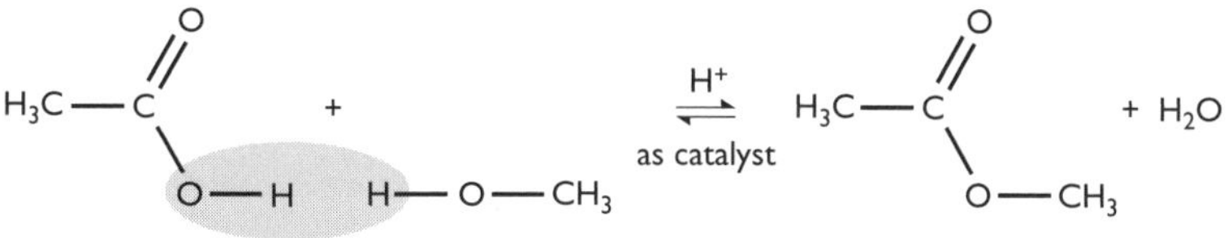

Ethanoic acid   Methanol   Methyl ethanoate   Water

or:

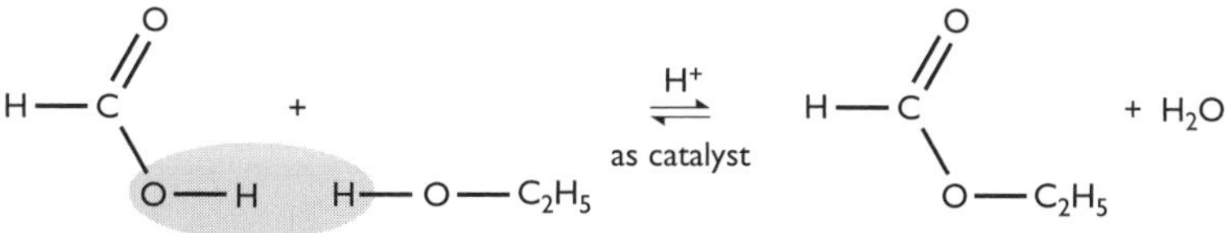

Methanoic acid   Ethanol   Ethyl methanoate   Water

Esters are used in flavourings and in perfumes. They often contribute to the flavour of fruits.

Pineapple

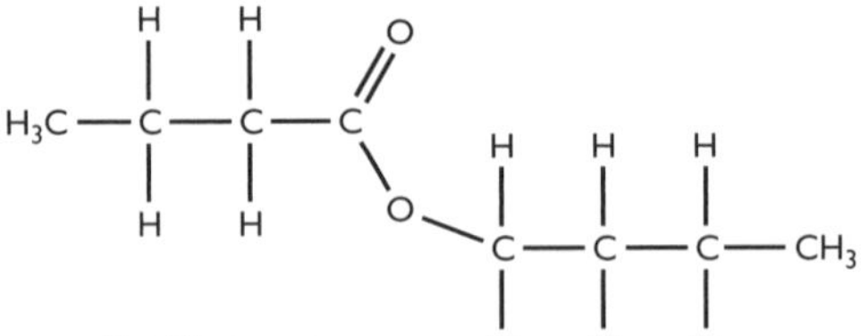

Butyl butanoate

Pear

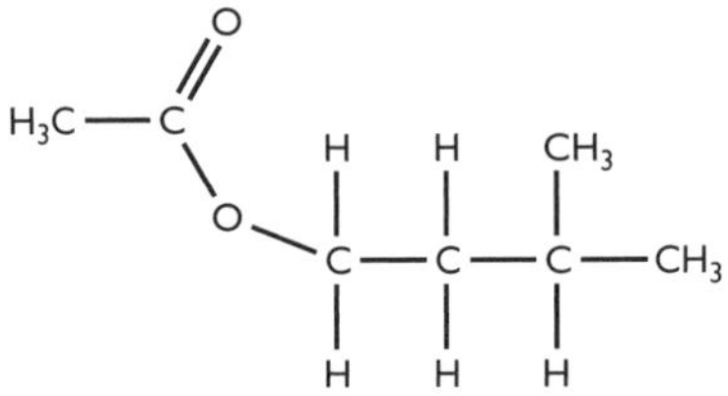

3-methylbutyl ethanoate

Apple

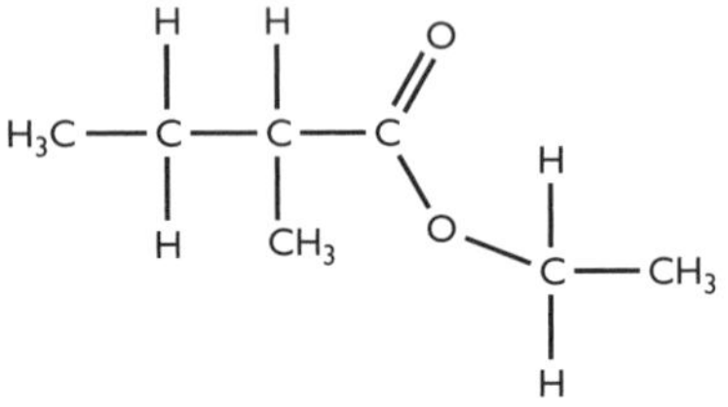

Ethyl 2-methylbutanoate

Esters, triglycerides, unsaturated and saturated fats

Esters can react with water. The **hydrolysis** reaction is slow and is carried out in the presence of either an acid ($H^+(aq)$) or a base ($OH^-(aq)$). **Acid-catalysed hydrolysis** leads to the formation of the carboxylic acid and the alcohol.

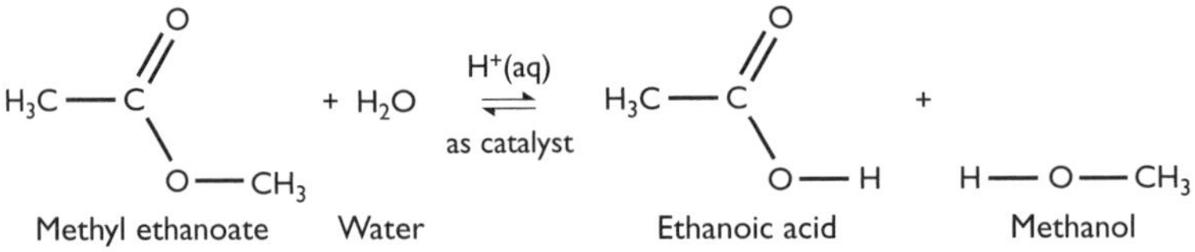

Methyl ethanoate   Water   Ethanoic acid   Methanol

Base-catalysed hydrolysis leads to the formation of the salt of the carboxylic acid (the carboxylate) and the alcohol.

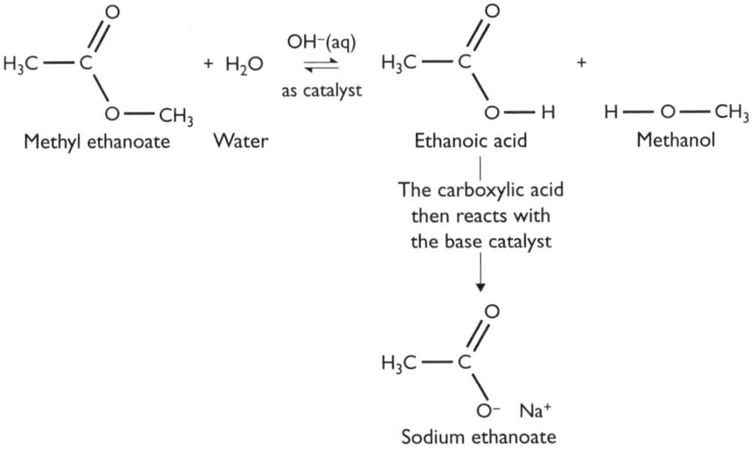

Methyl ethanoate   Water   Ethanoic acid   Methanol

The carboxylic acid then reacts with the base catalyst

Sodium ethanoate

Many esters occur naturally in plants and flowers, as well as in vegetable oils and in animal fats. The esters found in vegetable oils and animal fats are esters of fatty acids.

Common fatty acids are generally given trivial names, such as oleic acid (found in olive oil). Such names are in everyday use, but chemists know fatty acids by their systematic names. For example, oleic acid is (Z)-octadec-9-enoic acid, 18, 1(9). This sounds much more complicated than oleic acid, but the numbers at the end of the name show that the fatty acid contains 18 carbon atoms, with one Z (or *cis*) double bond on the ninth carbon (starting from the COOH end of the molecule). Thus, its formula is:

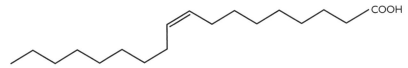

Fatty acids are an essential part of a healthy diet. They can be **saturated, mono-unsaturated** or **poly-unsaturated**. Saturated fatty acids (SFAs) contain no C=C double bonds and are essentially straight chains.

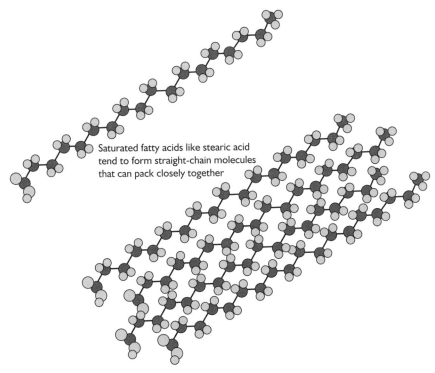

Saturated fatty acids like stearic acid tend to form straight-chain molecules that can pack closely together

The straight chains of SFAs can pack together closely, providing many points of contact and intermolecular forces. This increases viscosity and reduces volatility, making SFAs relatively dense and solid at room temperature.

Mono-unsaturated fatty acids (MUFAs) such as oleic acid ($CH_3(CH_2)_7CH=CH(CH_2)_7COOH$) have one C=C double bond. Oleic acid is found in most animal and vegetable fats that occur naturally as the Z (*cis*) isomer. The double bond creates kinks in the molecule, making it more difficult to stack unsaturated fatty acids. Hence they have fewer points of contact, fewer intermolecular forces and tend to be liquids at room temperature.

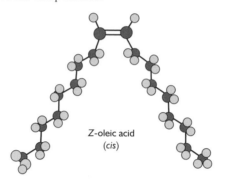

Z-oleic acid
(*cis*)

Poly-unsaturated fatty acids (PUFAs) contain more than one C=C double bond and hence have more than one kink. Therefore, poly-unsaturated fatty acids stay fluid even at low temperatures. PUFAs are synthesised by plankton and seaweed, which are eaten by fish and incorporated into their fatty tissue. This makes fish a ready source of poly-unsaturated fatty acids.

Naturally occurring fatty acids, found in both animals and plants, are mostly the Z (*cis*) form, and therefore have a distinctive *kinked* shape. Food manufacturers partially hydrogenate fatty acids (add hydrogen to some of the C=C double bonds) as it prolongs their shelf-life and makes them less likely to become rancid. The hydrogenation process converts the Z (*cis*) form into the E (*trans*) form, thus straightening out the kink in the molecule.

Exposure to prolonged heat, such as repeatedly using oil for deep frying, also creates E (*trans*) fatty acids. The chemical structures of the Z (*cis*) and E (*trans*) fats are identical, but the *geometrical* differences affect important biological processes.

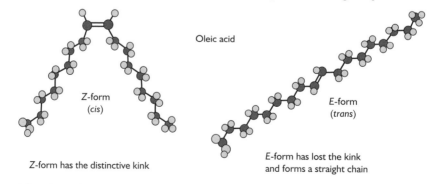

Oleic acid

Z-form
(*cis*)

E-form
(*trans*)

Z-form has the distinctive kink

E-form has lost the kink
and forms a straight chain

Straight-chain fatty acids (saturated fatty acids and E (*trans*) fatty acids) can stack together, leading to a build-up of plaque and thickening of the walls of the human arteries. Saturated fats, and possibly *E* (*trans*) fatty acids, can also lead to increased blood levels of cholesterol and low-density lipoproteins (LDLs). It has long been known that a diet high in saturated fat and cholesterol can lead to thickening of the walls of the arteries (atherosclerosis) and heart disease. For years we have been urged to lower the amount of saturated fat in our diets. Cholesterol is essential for the formation of cell membranes and several body hormones. It is insoluble in water and blood and is transported around the body by lipoproteins. The lipoproteins fall into two categories: high-density lipoprotein (HDL), which carries about one-third of the cholesterol and is known as **good cholesterol** as it carries cholesterol away from the arteries and back to the liver, and low-density lipoprotein (LDL), which is **bad cholesterol** that can lead to a build-up of plaque and to atherosclerosis. Recent studies show a correlation between the amount of *E* (*trans*) fatty acids and LDL. If people are worried about their blood cholesterol levels, they should avoid eating *E* (*trans*) fatty acids by using liquid vegetable oils.

Glycerides occur when one or more of these fatty acids become attached to a glycerol molecule. Glycerol contains three alcohol groups and is known as a triol. Its systematic name is propane-1,2,3-triol ($CH_2(OH)CH(OH)CH_2OH$). Glycerol can react with fatty acids to produce esters. There can be one fatty acid (monoglyceride), two fatty acids (diglycerides) or three fatty acids (triglyceride) attached to the glycerol.

Propane-1,2,3-triol
(glycerol)
backbone

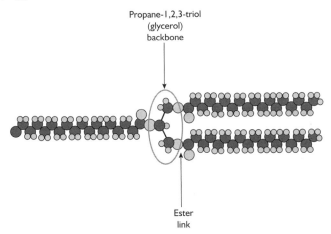

Ester
link

Naturally occurring fats and oils (triglycerides) can be hydrolysed by refluxing with a base. This produces propane-1,2,3-triol and the salts of the fatty acids. This is known as **saponification**. Saponification means 'the forming of soap' and modern soaps are made from blends of oils. The base hydrolysis of a triglyceride is shown below.

Vegetable oils and animal fats and their derivatives, especially methyl esters, are increasingly being used as alternative diesel fuels. They are known as **biodiesels**. A biodiesel can be defined as 'a mono-alkyl ester of long-chain fatty acids from renewable feedstocks such as vegetable oil or animal fats'. When a fuel is injected into the combustion chamber of a diesel engine, there is a time delay before ignition. This time delay is the basis of the **cetane number**. The shorter the time delay, the higher the cetane number. Hexadecane ($C_{16}H_{34}$) is a high-quality fuel with a short ignition time, and has been assigned a cetane number of 100. A diesel engine that is run on a fuel with a lower cetane number than its design specification will be harder to start, noisier, will operate roughly and have higher emissions. Biodiesels have higher cetane numbers than their hydrocarbon counterparts.

Amines

Basicity of amines

Like ammonia, primary amines are weak bases and will accept a proton from water to form an alkaline solution.

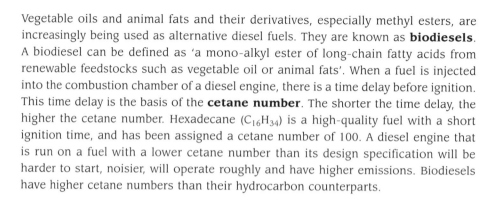

Ammonia and amines have a lone pair of electrons on the nitrogen atom. The lone pair accepts the proton and forms a dative covalent bond.

Ethyl groups have a *positive* **inductive effect** and push electrons along the bond towards neighbouring atoms. This increases the electron density on the nitrogen atom and makes ethylamine *more basic*.

If you are asked to define what is meant by the inductive effect, a suitable definition is: *the movement of electrons along a σ-bond*.

Phenyl groups have a *negative* inductive effect and the lone pair of electrons on the nitrogen is incorporated (delocalised) into the ring. This decreases the electron density on the nitrogen atom and makes phenylamine *less basic*.

The order of basicity is: ethylamine > ammonia > phenylamine.

Primary amines are weak bases and, therefore, react with acids to produce salts.

$C_2H_5NH_2 + HCl \rightarrow C_2H_5NH_3{}^+Cl^-$

Ethylamine

$C_6H_5NH_2 + HCl \rightarrow C_6H_5NH_3{}^+Cl^-$

Phenylamine

Preparation of amines

Primary aliphatic amines such as ethylamine ($CH_3CH_2NH_2$) can be prepared from the reaction between a halogenoalkane and ammonia:

$CH_3CH_2Cl + NH_3 \rightarrow CH_3CH_2NH_2 + HCl$

Primary aryl (or aromatic) amines such as phenylamine ($C_6H_5NH_2$) can be prepared by the reduction of nitrobenzene. Nitrobenzene is refluxed with the reducing agent: tin and concentrated hydrochloric acid. The equation for this reaction is shown below. [H] represents the reducing agent.

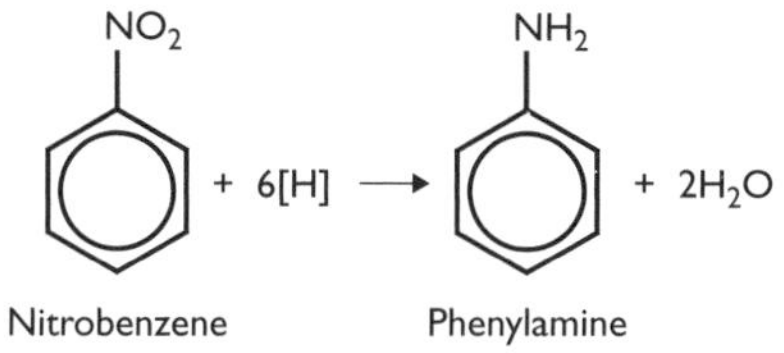

Nitrobenzene   Phenylamine

Synthesis of azo dyes

Aromatic amines, like phenylamine, are essential for the manufacture of azo dyes. The synthesis involves two stages:
- formation of a diazonium compound
- a coupling reaction with a phenol to form the azo dye

Stage I

Reagents: nitrous acid (HNO_2) formed from $NaNO_2$ and excess HCl

Conditions: temperature below 10°C

Balanced equation:

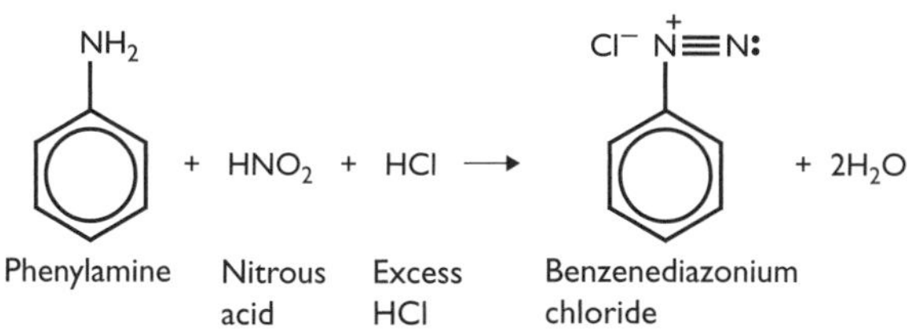

Phenylamine   Nitrous   Excess   Benzenediazonium
           acid   HCl   chloride

The reaction has to be kept below 10°C because benzenediazonium chloride is unstable and reacts readily with water to produce phenol, N_2 and HCl.

Stage 2

Reagent: phenol

Conditions: alkaline solution (in the presence of OH⁻(aq))

Balanced equation:

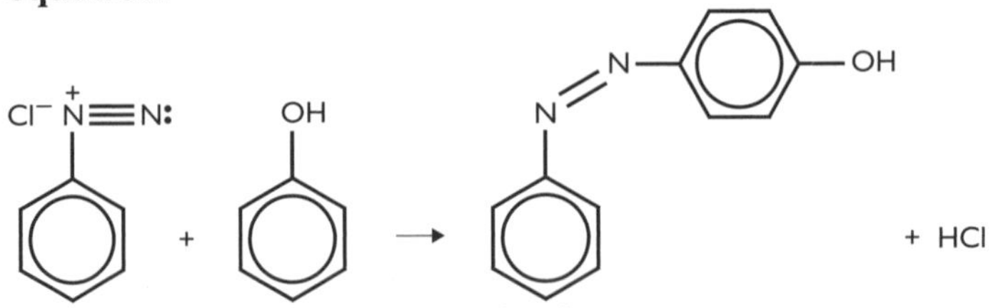

Benzenediazonium chloride   Phenol   Azo dye   + HCl

The $-N=N-$ group absorbs light, making azo compounds brightly coloured. The exact colour depends on the other substituents on the aromatic rings.

Module 2: Polymers and synthesis

Amino acids, proteins and chirality

Amino acids

The general formula of an α-amino acid is $RCH(NH_2)COOH$, where R represents the side-chain.

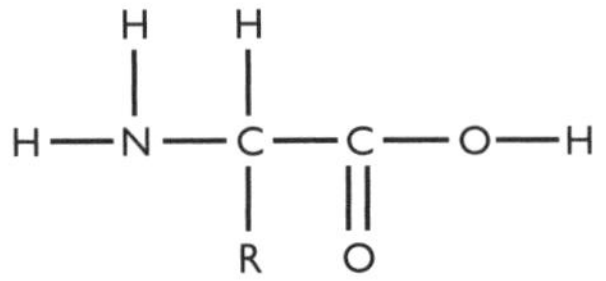

In an α-amino acid, the amine and the carboxylic acid groups are both attached to the same C atom.

The simplest α-amino acid is aminoethanoic acid, or **glycine**, in which the R group is H. In 2-aminopropanoic acid, or **alanine**, the R group is CH_3. These two amino acids are shown below. Alanine has two optical isomers. Glycine is not optically active. Alanine has an asymmetric centre (∗) and glycine doesn't.

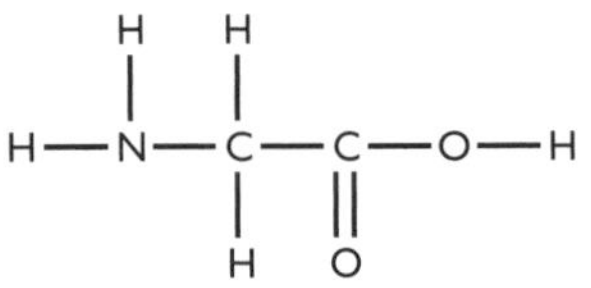

Glycine 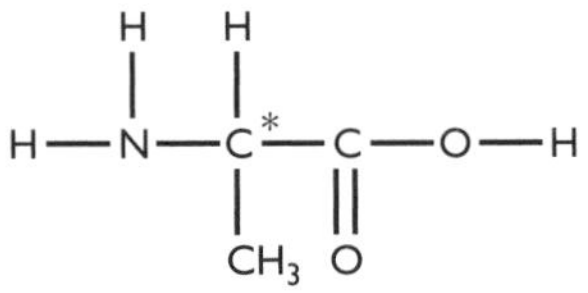   Alanine

An asymmetrical carbon atom is a carbon atom bonded to four different atoms or groups. Glycine is not optically active because the carbon atom is bonded to two hydrogen atoms. The optical isomers of alanine are shown below.

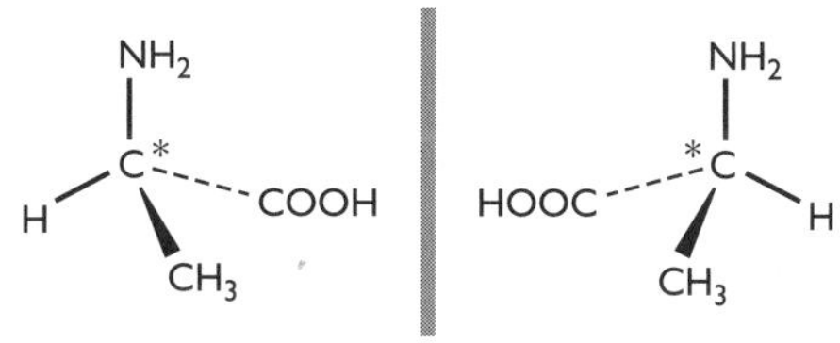

Amino acids are bi-functional because they contain two functional groups — carboxylic acid and amine.

Behaving as a carboxylic acid

An amino acid can react with a base to produce a salt:

$$NH_2CH_2COOH(aq) + NaOH(aq) \rightarrow NH_2CH_2COO^-Na^+(aq) + H_2O(l)$$

Glycine

An amino acid can react with an alcohol to produce an ester:

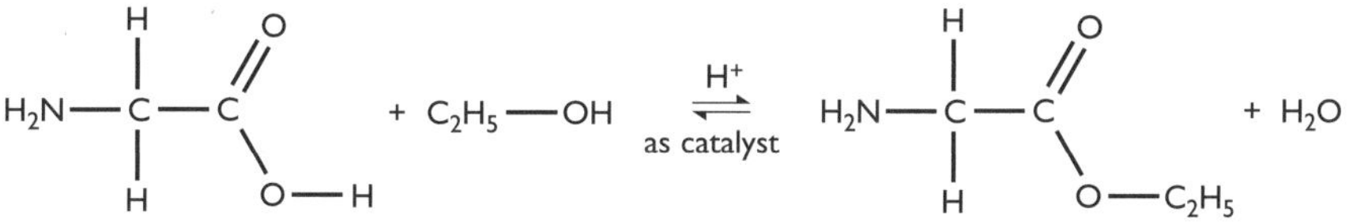

Behaving as a primary amine

An amino acid can also behave as a primary amine and react with an acid to produce a salt:

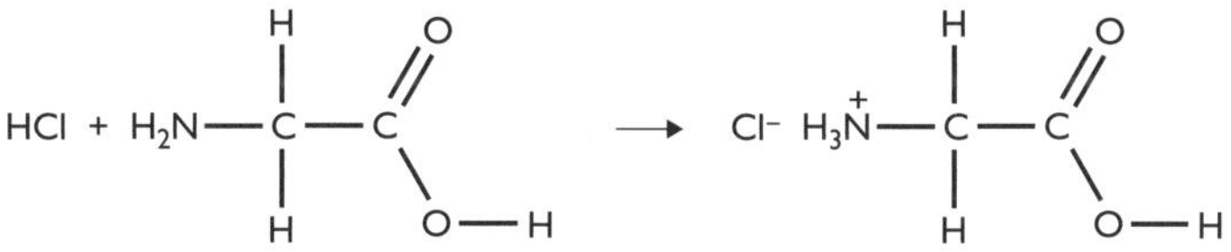

Properties dependent on both functional groups

Amino acids also display properties that depend on both functional groups. Unlike most organic compounds, amino acids tend to have very high melting points and are water-soluble. This is due to the formation of **zwitterions**.

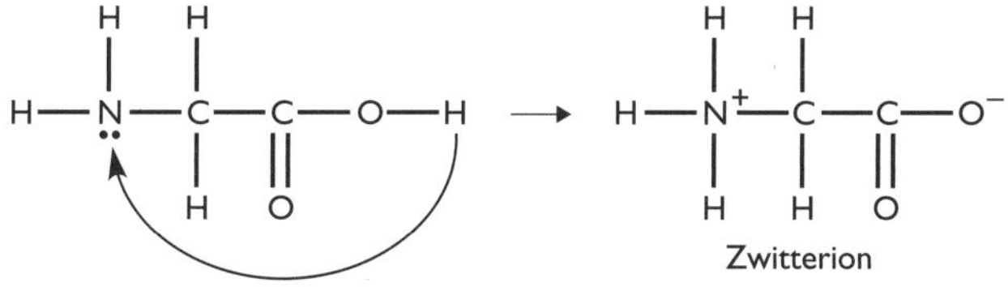

Zwitterion

The zwitterion, for each amino acid, exists at a particular pH. The pH at which a zwitterion exists is known as the **isoelectric point**. If the amino acid is in an acidic solution, it forms a cation. If it is in an alkaline solution, it forms an anion.

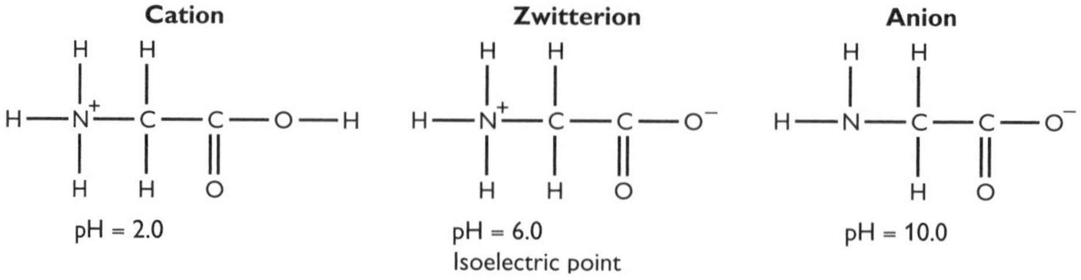

Cation — pH = 2.0   Zwitterion — pH = 6.0, Isoelectric point   Anion — pH = 10.0

Amino acids can react to form peptides.

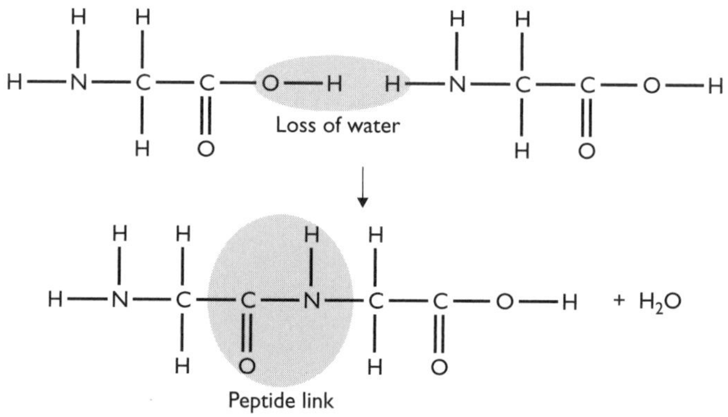

Loss of water

Peptide link + H_2O

If two different amino acids, such as glycine (Gly), and alanine (Ala) react, it is possible to form different dipeptides:

Gly + Ala → Gly—Ala

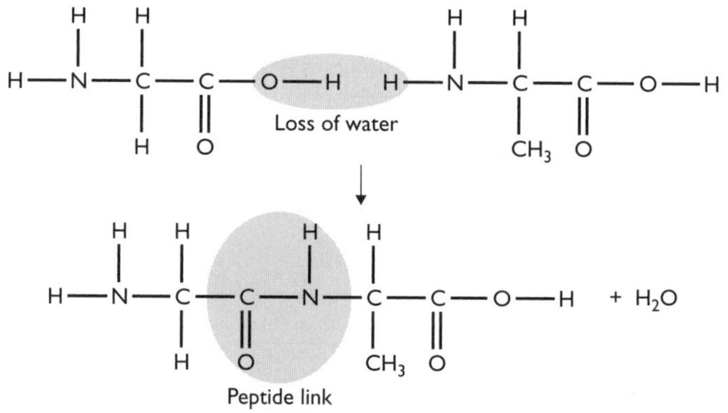

Loss of water

Peptide link + H_2O

Ala + Gly → Ala—Gly

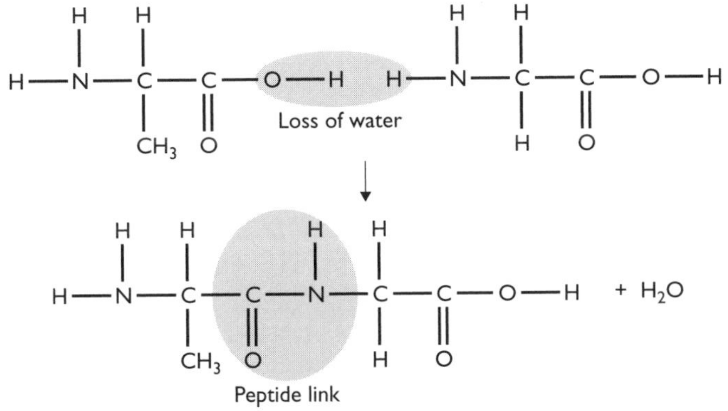

Loss of water

Peptide link

All dipeptides can react further with other amino acids, extending the chain length. This leads to the formation of polypeptides and proteins.

Peptides and proteins can be hydrolysed back to the amino acids by refluxing the peptide (or protein) with $6.0\,mol\,dm^{-3}$ hydrochloric acid or with aqueous sodium hydroxide solution. You are expected to be able to identify the amino acids present in a polypeptide.

Acid hydrolysis produces the cation:

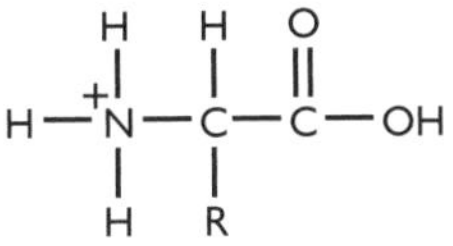

cation

Alkaline hydrolysis produces the anion:

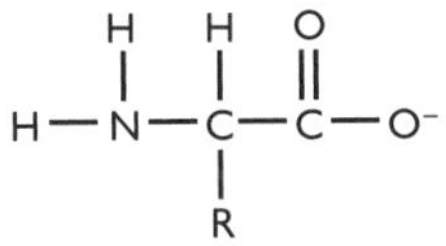

anion

Polyesters and polyamides

There are two categories of polymer: addition polymers and condensation polymers.

Addition polymers

Alkenes can undergo an addition reaction in which one alkene molecule joins to others and a long molecular chain is built up. The individual alkene molecule is referred to as a **monomer** and the long-chain molecule is a **polymer**.

Polymerisation can be initiated in a variety of ways and the initiator is often incorporated at the start of the long molecular chain. However, if the initiator is disregarded, the empirical formulae of the monomer and the polymer are the same. Common monomers and the polymers formed from them include:

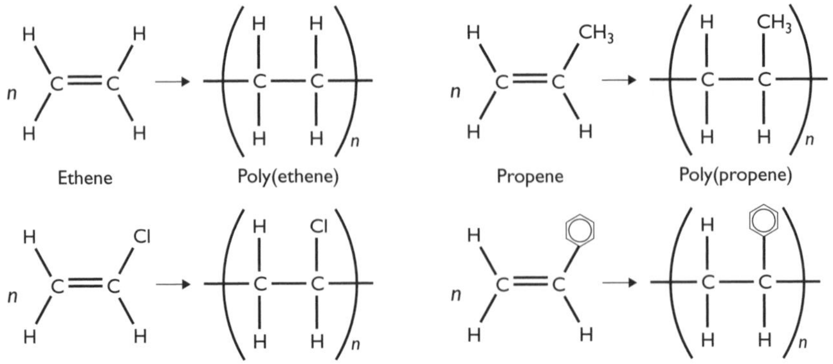

| Ethene | Poly(ethene) | Propene | Poly(propene) |

| Chloroethene or vinyl chloride | Poly(chloroethene) or PVC | Phenylethene (styrene) | Polystyrene |

The bonds in addition polymers are strong, covalent and non-polar, which makes most polymers resistant to chemical attack. Also, they are **non-biodegradable** because they are not broken down by bacteria. The widespread use of these polymers has created a major disposal problem.

Condensation polymers

Condensation polymers are formed when monomers react together and 'condense' out a small molecule such as H_2O or HCl. There are two main types: polyesters and polyamides.

Polyesters

Terylene is a common polyester made by the reaction between the monomers ethane-1,2-diol and benzene-1,4-dicarboxylic acid.

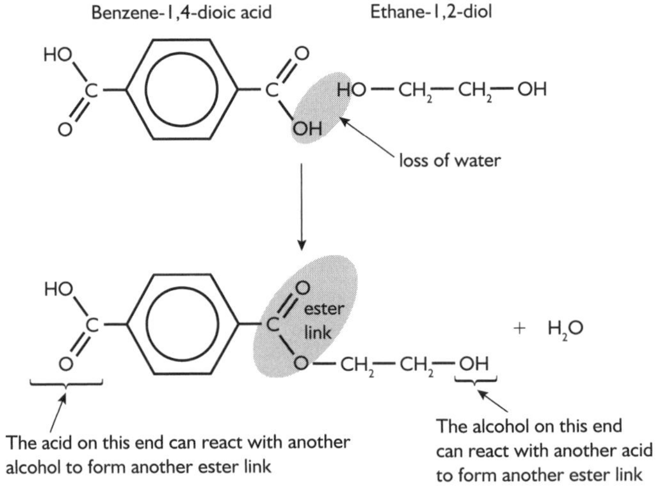

Benzene-1,4-dioic acid   Ethane-1,2-diol

loss of water

ester link

$+$ H_2O

The acid on this end can react with another alcohol to form another ester link

The alcohol on this end can react with another acid to form another ester link

The resulting polymer is almost linear. This means that the polymer chains can be packed closely together. The close packing produces strong intermolecular forces that enable the polymer to be spun into thread.

Polyamides

Polyamides are prepared from two monomers, one with an amine group at each end and the other with a carboxylic acid at each end. Nylon-6,6 is made from the two monomers 1,6-diaminohexane and hexane-1,6-dicarboxylic acid. It is called nylon-6,6 because each monomer contains six carbon atoms.

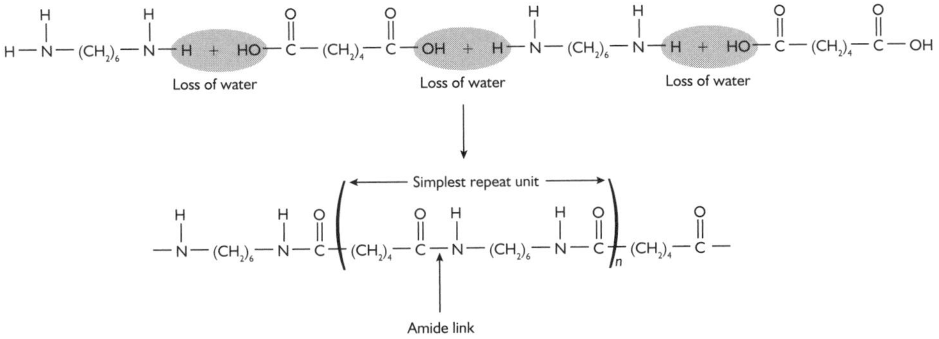

Loss of water   Loss of water   Loss of water

Simplest repeat unit

Amide link

Nylon forms a strong, flexible fibre when it is melt-spun.

Kevlar is another polyamide. It is stronger than steel and is fire-resistant. It is used for making bulletproof vests, crash helmets and protective clothing used by fire fighters. It is made from the two monomers benzene-1,4-diamine and benzene-1,4-dicarboxylic acid.

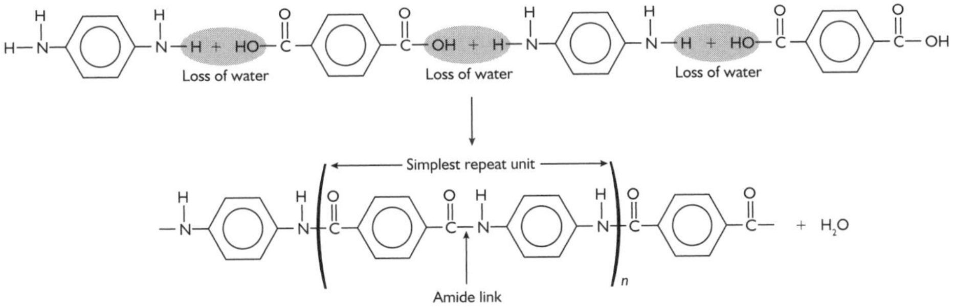

Loss of water   Loss of water   Loss of water

Simplest repeat unit $+ H_2O$

Amide link

Hydrolysis and degradable polymers

The ester link in a polyester and the amide link in a polyamide are both polar links and are subject to acid- and base-catalysed hydrolysis.

Ester link   Amide link

Acid hydrolysis of a polyester results in the formation of a diol and a dioic acid:

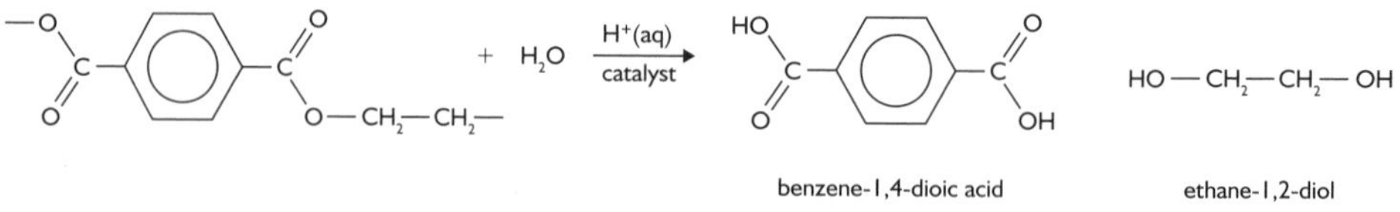

benzene-1,4-dioic acid   ethane-1,2-diol

Base hydrolysis of a polyester also forms a diol. However, the dioic acid formed reacts with the base catalyst to produce the dioate salt. The products of refluxing terylene with an aqueous solution of NaOH(aq) are:

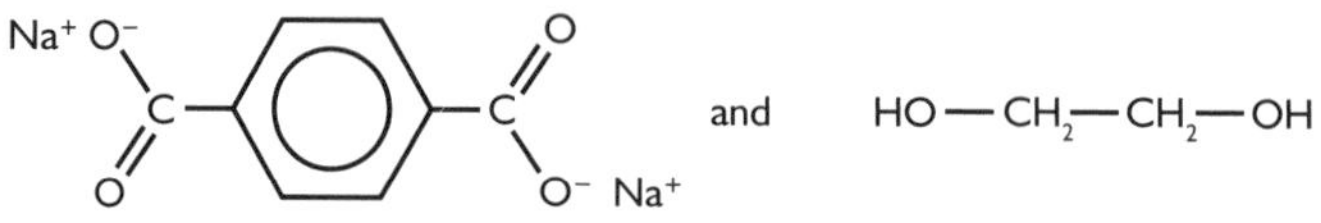

Polyamides hydrolyse in a similar way, but acid-catalysed hydrolysis forms the dioic acid and the di-salt of the diamine, while base-catalysed hydrolysis forms the diamine and the dioate salt of the dioic acid. This is summarised in the reaction scheme below:

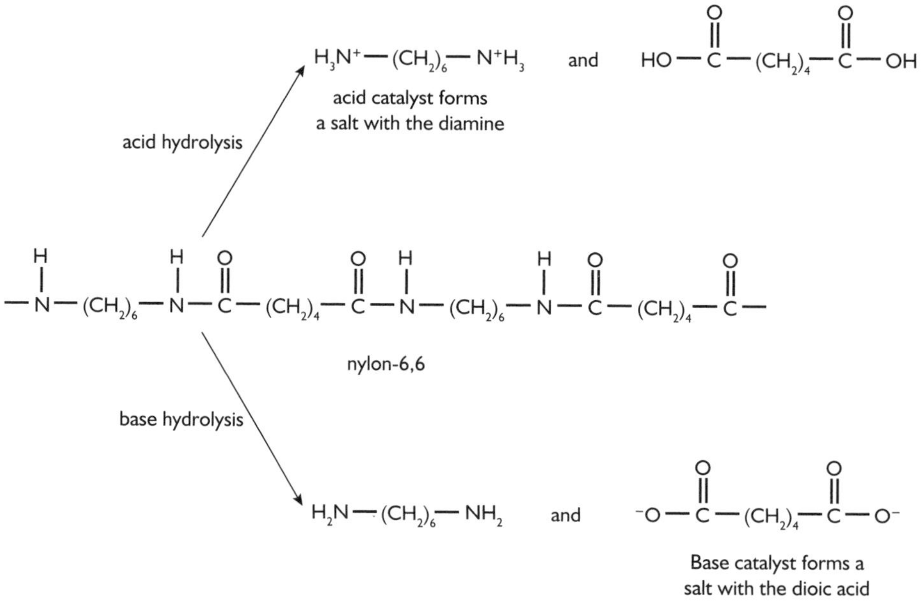

acid hydrolysis

$H_3N^+—(CH_2)_6—N^+H_3$ and $HO—\overset{O}{\overset{||}{C}}—(CH_2)_4—\overset{O}{\overset{||}{C}}—OH$

acid catalyst forms
a salt with the diamine

nylon-6,6

base hydrolysis

$H_2N—(CH_2)_6—NH_2$ and $^-O—\overset{O}{\overset{||}{C}}—(CH_2)_4—\overset{O}{\overset{||}{C}}—O^-$

Base catalyst forms a
salt with the dioic acid

Chemists are aware of the environmental impact of using compounds derived from fossil fuels and the problems associated with the disposal of plastic waste. Condensation polymers can be broken down by hydrolysis, and chemists are developing polymers that are photodegradable, such as poly(lactic acid) (PLA). This is prepared from 2-hydroxypropanoic acid (also known as lactic acid, $CH_3CH(OH)COOH$). Lactic acid contains both the alcohol group (—OH) and the carboxylic acid group (—COOH), which can react together to produce ester linkages.

The introduction of polymers such as PLA reflects the role of chemists in minimising environmental impacts by using renewable sources and by developing degradable polymers. PLA is particularly attractive as a sustainable alternative to petrochemical-derived products, since the monomer can be produced from corn starch or sugar cane, which are renewable feedstocks. PLA acid is fully compostable, degrading to form CO_2 and H_2O in a relatively short time.

PLA is more expensive than many petroleum-derived plastics, but its price has been falling as more is being manufactured. It is uncertain how far the price will fall, and the degree to which PLA will be able to compete with non-sustainable petroleum-derived polymers.

Synthesis

It is often not possible to convert one chemical into another using a single reaction. Very often intermediate compounds have to be formed. The flow chart of aliphatic chemistry below covers the functional groups presented in this unit. It shows the links between the functional groups, as well as giving the essential reagents and conditions for reaction.

Flow chart 1

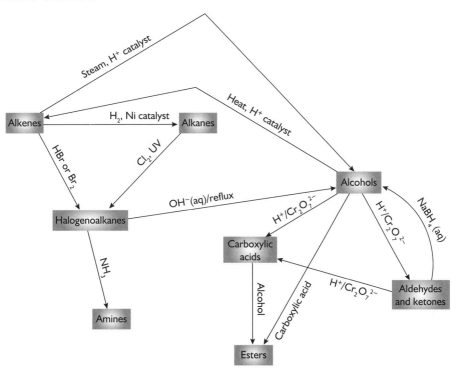

Using the flow chart, it can be seen that an alkene can be converted into an alkane in a single step. However, the reverse conversion cannot be achieved in a single step. To convert an alkane into an alkene, the alkane is first converted into a halogenoalkane, then into an alcohol, and finally into an alkene. This multi-stage conversion is less efficient than a single-stage conversion. You can use the flow chart to work out lots of other multi-stage conversions.

Flow chart 2 is much simpler and links together all of the aromatic chemistry you need to know.

Flow chart 2

The OCR specification requires that you should be able to work out multi-stage syntheses for preparing organic compounds. This involves knowledge of all the reactions in the specification, but you may be provided with unknown reactions that you have to use.

Chirality in pharmaceutical synthesis

Chemicals with a chiral centre can exist as one of two optical isomers. Optical isomers are so called because they behave differently in the presence of plane-polarised light. One isomer will rotate the plane-polarised light to the right (the D isomer) and the other will rotate it to the left (the L isomer). Each isomer is optically active. The D- and the L-forms have different shapes. Many biochemical processes require molecules of a specific shape, so it is easy to see why one isomer is naturally predominant. All naturally occurring α-amino acids occur in the L-form only.

The preparation of a single chiral compound in the laboratory is extremely difficult. Laboratory synthesis is very likely to result in the formation of equal amounts of both optical isomers. The synthetic mixture would probably contain 50% of the isomer that rotates plane-polarised light to the right (the D-isomer) and 50% of the isomer that rotates plane-polarised light to the left (the L-isomer). It may be that only one of the isomers has the correct shape and is pharmaceutically active. The other isomer could have adverse side-effects or may make the correct isomer less effective. Therefore, it is vital that a pharmaceutical manufacturer isolates the effective isomer. Separation of optical isomers can be achieved, but it is tedious and expensive. Techniques such as chiral chromatography can be used.

Chromatography is discussed in more detail in the next section. Essentially, normal chromatography cannot distinguish between two optical isomers. However, if the column is packed with a solid that contains an active site in the form of either an enzyme or a chiral **stationary phase**, the column absorbs one optical isomer more than the other, allowing separation.

By contrast, synthesis using naturally occurring enzymes or bacteria results in the formation of a single optical isomer. Enzymes and bacteria achieve this because the active sites have a specific shape that promotes certain reactions. In the case of medicines, the pharmaceutically active isomer is produced, eliminating possible adverse side-effects from the other optical isomer and also avoiding the need to separate the pharmacologically active isomer from the inactive isomer.

Nowadays, single optical isomers are also produced by using starting materials such as L-amino acids or L-sugars, or a chiral catalyst. Chiral catalysts are used to develop new drugs and to produce flavourings, sweetening agents and insecticides. They also have wide applications in material and medical science.

Module 3: Analysis

Chromatography

Chromatography is a small-scale analytical technique that separates components in a mixture. All types of chromatography contain a **stationary** and a **mobile** phase. Different types of chromatography separate the components in a mixture by either **adsorption** or **partition**.

Thin-layer chromatography (TLC)

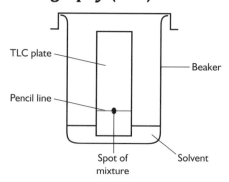

The TLC plate can be made of glass, metal or plastic. It is covered with a uniform, thin layer of either silica gel (SiO_2) or alumina (Al_2O_3), which is the stationary phase. The mixture is spotted on to the base line (which is drawn in pencil), and is allowed to dry. The plate is placed in the solvent (the mobile phase) and the beaker is covered with a watch glass to ensure that the beaker is saturated with solvent vapour. This prevents the solvent from evaporating as it rises up the plate.

The components in the mobile phase travel at different rates because they interact with the surface of the stationary phase. This is called **adsorption**. The surface of the stationary phase is polar, leading to the creation of hydrogen bonds and van der Waals forces, and to dipole–dipole interactions with components in the mixture. The greater the interaction, the greater the adsorption, thus restricting the distance travelled by that component.

As the solvent moves up the plate, the components in the mixture also rise with it. The speed at which each component rises depends on the solubility of the component in the solvent and on how much the component sticks (adsorbs) to the stationary phase. The latter depends on the polarity of the component and its ability to form hydrogen bonds or van der Waals forces with the surface of the stationary phase.

TLC achieves separation by adsorption. The components in a mixture are often identified by using R_f values. The R_f value stands for **retardation factor** and is measured by using the equation:

$$R_f = \frac{\text{Distance moved by spot/solute}}{\text{Distance moved by solvent}}$$

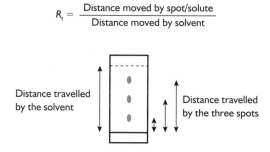

Gas chromatography (GC)

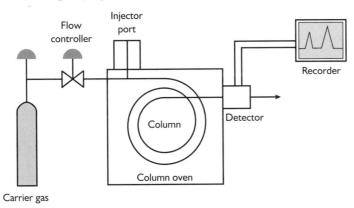

Gas chromatography involves a sample being vaporised and injected into the chromatographic column. The mobile phase is an unreactive or an inert gas such as nitrogen or one of the noble gases. The gas is known as the **carrier gas** and flows under pressure through the column.

The stationary phase is either a liquid or a solid adsorbed onto the surface of an inert solid. The way in which separation is achieved depends on whether the stationary phase is a liquid or a solid.

If the stationary phase is a liquid on the surface of the inert solid, separation depends on the relative solubility of the component in the stationary and the mobile phase, and separation is achieved by **partition**.

Separation by partition is achieved because solutes are not equally soluble in the mobile and the stationary phases, meaning that an equilibrium is set up between the mobile and the stationary phase.

$$\frac{\text{Concentration of solute in mobile phase}}{\text{Concentration of solute in stationary phase}} = \text{Constant}$$

If the stationary phase is a solid on the surface of the inert solid, then separation depends on the adsorption of the component onto the stationary phase, and separation is achieved by **adsorption**.

The time between injection of the sample and the emergence of a component from the column is called the retention time. For examination purposes, retention time is defined as *the time from the injection of the sample for each component to leave the column*.

Retention time depends on the volatility of the solute and the relative solubility of the solutes in the mobile and the stationary phases.

A detector is used to monitor the outlet stream from the column, to determine the time at which each component reaches the outlet and the amount of that component. The substances are identified qualitatively by the order in which they emerge from the column and quantitatively by the area of the peak of each component. The recorder produces a **chromatogram** showing each component as a separate peak.

Analysis by GC has its limitations because similar compounds often have very similar retention times. For example, it is highly likely that in a mixture of natural gases, such as methane, ethane, propane and butane, some of the peaks would overlap. Identification of unknown compounds is difficult, because reference times vary depending on the flow rate of the carrier gas and on the temperature of the column. Retention times can even vary from one GC machine to another. These limitations have largely been overcome by coupling together GC with a mass spectrometer.

GC–MS

The combination of gas chromatography with mass spectrometry (GC–MS) provides a powerful analytical tool that is widely used in areas such forensic science, environmental analysis and airport security.

The components in a mixture are separated by gas chromatography and each component is then placed separately into a mass spectrometer for analysis. The component are vaporised and then ionised. In Unit F322, the process of fragmentation was outlined and a number of simple fragment ions were identified in order to determine the relative isotopic masses of the elements. For example, the mass spectrum of a sample of propane looks like this:

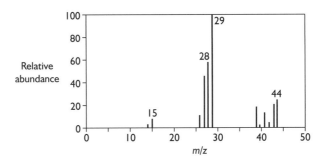

Propane has relative molecular mass of 44 and, as expected, the peak furthest to the right of the spectrum represents the ion $C_3H_8^+$. This is called the **molecular-ion peak**. However, the bombardment of electrons also creates ions from fragments of the molecule and these too are registered on the printout. Some examples have been labelled on the spectrum.

The peak at 29 occurs because a CH_3 unit has been broken from the $CH_3CH_2CH_3$ chain and the ion $CH_3CH_2^+$ has been detected.

$$\left(H_3C - CH_2 \!\mid\! CH_3\right)^+_{(g)} \longrightarrow \left(H_3C - CH_2\right)^+_{(g)} + CH_3\,(g)$$

Similarly, the peak at 15 represents a CH_3^+ ion.

$$\left(H_3C-CH_2\!\!-\!\!CH_3\right)^+_{(g)} \longrightarrow CH_3^+(g) \ + \ \left(H_3C-CH_2\right)_{(g)}$$

and it is possible to suggest the identity of all other peaks in the spectrum.

Even in molecules as simple as propane there are a large number of lines. The combination of the lines and their height are specific to that individual compound, meaning that the fragmentation peaks are often said to be a fingerprint of the molecule. This fingerprint can then be cross-matched against a large computer database to identify any specific compound.

Coupling together GC and MS enables separation of the components in a mixture (GC) and the identification of each component (MS). Together the two techniques are widely used in many scientific fields, such as forensics, environmental analysis, airport security, and food and drink analysis, as well as being used extensively in medical applications.

Spectroscopy

NMR spectroscopy

NMR spectroscopy involves interaction of atomic nuclei with radio waves that are at the low-energy end of the electromagnetic spectrum.

If the nucleus of an atom contains an odd number of nucleons (protons and neutrons), it has a net nuclear spin. The nucleus behaves like a tiny bar magnet, and generates a magnetic moment as it spins. Adjacent nuclei also have magnetic moments, meaning that each nucleus is affected by its neighbouring nuclei. The frequency at which each nucleus absorbs radio waves depends on the surrounding atoms.

If the nucleus of an atom contains an even number of nucleons, the nucleus does *not* have a net nuclear spin and *cannot* be detected using radio frequency. For example, ^{12}C and ^{16}O cannot be detected.

If the nucleus of an atom contains an odd number of nucleons, the nucleus will interact with radio waves. ^1H and ^{13}C both absorb energy in the radio-wave part of the spectrum, but the frequency of the absorption varies depending on the surrounding atoms, such that the exact frequency absorbed depends on chemical environment. This variation in the frequency absorbed is key to the determination of the structure, and is known as the **chemical shift** (δ). All absorptions are measured relative to tetramethylsilane (TMS), whose chemical shift δ is set at zero. TMS is used as a standard because:

- it is chemically inert and does not react with the sample
- it is volatile and easy to remove afterwards
- it absorbs at a higher frequency than other organic compounds, so that it does not overlap with the sample

All other peaks are measured relative to TMS.

You do not have to learn the chemical shift values of any of these peaks. All relevant absorptions for ^1H and ^{13}C NMR are listed in the data booklet, which you will be given in the examination. However, it is essential that you are able to recognise different chemical environments.

Carbon-13 NMR spectroscopy

Carbon-12 (^{12}C) is the most abundant isotope of carbon, but it does not have spin because it has an even number of protons and an even number of neutrons. The second principal isotope of carbon, carbon-13 (^{13}C), can be detected using low-energy radio waves, and it is possible to generate ^{13}C NMR spectra. The ^{13}C atom is about 6000 times more difficult to detect than ^1H atoms, because of its low abundance (only about 1.1% of naturally occurring carbon is ^{13}C) and its low magnetic moment. Interaction between adjacent ^{13}C atoms is highly unlikely because of their low abundance. The ^{13}C atoms do interact with adjacent protons, but these interactions are removed, so that all absorptions appear as singlets.

All ^{13}C NMR spectra are decoupled, so that all peaks appear as singlets. Each peak represents a different carbon atom environment.

Ethanol has two carbons: C_1 and C_2, which means that there will be two separate peaks — one for each carbon environment.

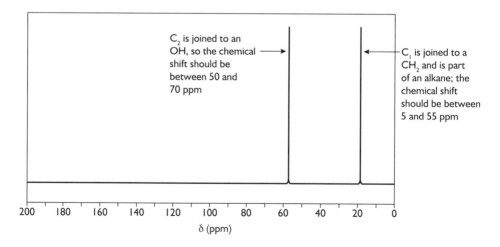

The key to interpreting ^{13}C NMR is to identify the number of different carbon environments and match them up to the groups in the data sheet.

Example 1

Determine the number of carbon environments in propan-2-ol. For each environment, predict the chemical shift, i.e. the δ value.

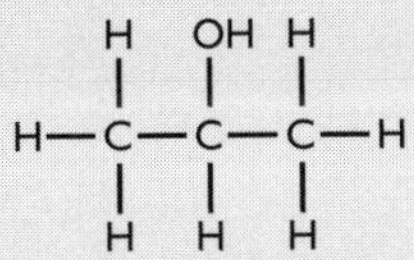

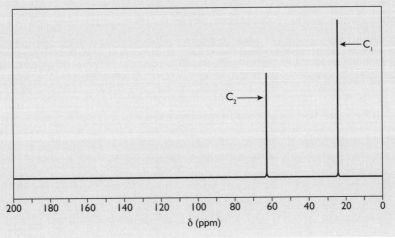

There are two different C environments: C_1 and C_2

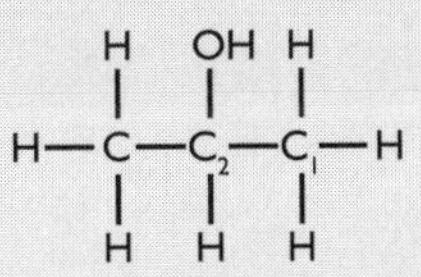

The ^{13}C NMR should therefore contain two peaks:

- C_2 is next to an OH group and should therefore have a δ value between 50 and 70 ppm.
- C_1 is part of an alkyl group and should therefore have a δ value between 5 and 55 ppm.
- However, because the adjacent C is bonded to an OH it will be towards the high end of the range.

The ^{13}C spectrum of propan-2-ol is shown below.

Proton NMR spectroscopy

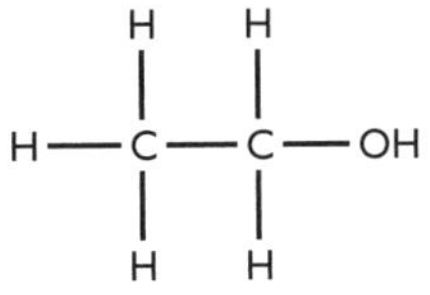

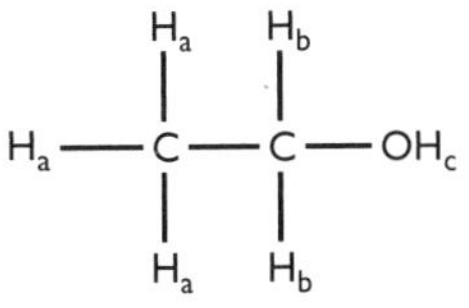

Ethanol (C_2H_5OH) contains six H atoms, but they are not all identical.

The three hydrogens in the CH_3 group are all in the same environment and can be labelled H_a; so are the hydrogens in the middle CH_2 group (labelled H_b). The H in the OH group is different from all the rest (labelled H_c). In ^1H (proton) NMR this leads to three different absorptions and hence three different peaks.

The six hydrogens in ethanol are in three different environments, so the NMR spectrum of ethanol is expected to contain three peaks with different chemical shifts.

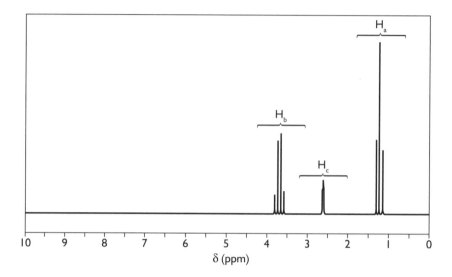

In the NMR spectrum of ethanol, each peak is a different size, and some of the peaks (H_a and H_b) are split. The relative size of each peak reflects the number of hydrogens in each environment:

- H_a — there are three hydrogens in this environment.
- H_b — there are two hydrogens in this environment.
- H_c — there is one hydrogen in this environment.

It follows that the relative intensity of each peak, $H_a:H_b:H_c$, is 3:2:1.

The hydrogen atoms attached to one carbon atom influence the H atoms on adjacent carbon atoms. This is called **spin–spin coupling**. The easiest way to predict the splitting pattern is to count the number of H atoms on the adjacent carbon atoms and use the '$n + 1$' rule, where n is the number of hydrogen atoms on the adjacent carbon atoms.

In the NMR spectrum of ethanol, each of the peaks is split differently.

- H_a is next to two hydrogens in the CH_2 group and hence is split into (2 + 1) — a **triplet**.
- H_b is next to three hydrogens in the CH_3 and hence is split into (3 + 1) — a **quartet**.
- H_c is not attached to a carbon atom and hence does not undergo spin–spin coupling. It is, therefore, a **singlet**.

Hence, the high-resolution NMR spectrum of ethanol is expected to have three peaks of relative intensity 3:2:1, split into a triplet, a quartet and a singlet.

The exact position (the chemical shift) of each peak can be obtained from the data sheet that is supplied in the examination.

- In the ^1H (proton) NMR of ethanol, the CH_3 (H_a) has a chemical shift between 0.7 and 1.6 ppm.
- The CH_2 (H_b) has a chemical shift between 3.3 and 4.3 ppm.
- The —OH (H_c) has a chemical shift between 1.0 and 5.5 ppm.

Use of D₂O

The O—**H** and the N—**H** peaks have chemical shifts that are variable and sometimes outside the range 1.0–5.5 ppm, so they can be difficult to assign. When alcohols, carboxylic acids or amines are dissolved in water, there is a rapid exchange between the protons in the functional groups (known as **labile protons**) and the protons in the water.

If water is replaced with deuterated water (^2H_2O), the peak at H_c disappears. The H_c proton is replaced by deuterium (^2H), which does not absorb in this region of the spectrum.

Deuterated water can be written as D_2O. If ethanol is dissolved in deuterated water, then:

This is no longer detected and
the peak for H_c disappears

The use of ^2H_2O to identify labile protons is a valuable technique in proton (^1H) NMR.

When samples are prepared for NMR, it is often necessary to dissolve them in a suitable solvent. Solvents containing hydrogen are clearly unsuitable, as the hydrogen atoms in the solvent would also be detected and thus interfere with the spectrum. This is overcome by using deuterated solvents such as $CDCl_3$.

Example 2

Draw propan-1-ol and determine:
- the number of different hydrogen environments
- the relative ratio numbers of the peaks
- the splitting of each peak
- the chemical shift of each peak

	Number of peaks	Ratio numbers	Splitting	Chemical shifts δ (ppm)
$H_a - C - C - C - OH_d$ with H_a, H_b, H_c on top and H_a, H_b, H_c on bottom	Four peaks H_a, H_b, H_c, H_d	H_a, H_b, H_c, H_d $3:2:2:1$	H_a – triplet H_b – sextet H_c – triplet H_d – singlet	H_a 0.7–1.6 H_b 1.2–1.4 H_c 3.3–4.3 H_d 1.0–5.5* *OH peaks are very variable and should be confirmed by using D_2O as the solvent

Example 3

Compound **A** has an empirical formula C_2H_4O and a molar mass of 88. The ^1H NMR of compound **A** is shown below. Deduce the identity of compound **A**.

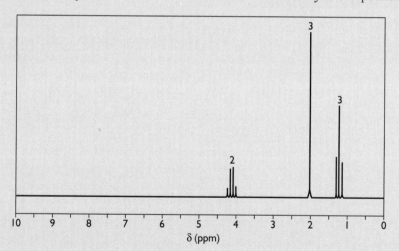

δ (ppm)

There are many different ways to solve this. One method is shown below:
- C_2H_4O has a molar mass = 24 + 4 + 16 = 44 g mol^{-1}. Therefore, the molecular formula is $C_4H_8O_2$.
- The NMR shows three peaks, so there are three different H environments.
- The ratio numbers 3:3:2 add up to eight, which indicates that the molecule is likely to contain two CH_3 groups and a CH_2.

- The peak at 1.2 is a triplet, indicating that it is next to a CH_2.
- The peak at 4.2 is a quartet, indicating that it is next to a CH_3. This indicates the presence of a CH_3-CH_2- grouping.
- The peak at 2.1 is a singlet, suggesting that it is a CH_3-, and the adjacent C has no hydrogens. The data sheet confirms that CH_3- bonded to a carbonyl (C=O) absorbs in this region.
- Compound A is known to contain CH_3-CH_2- and

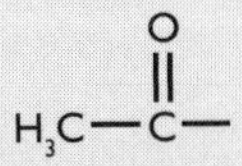

$$H_3C-C(=O)-$$

- The molecular formula is $C_4H_8O_2$, so the extra oxygen atom is likely to be part of an ester group.

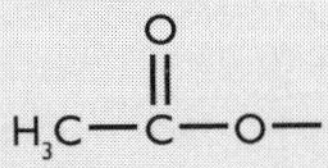

$$H_3C-C(=O)-O-$$

- Therefore Compound A is ethyl ethanoate.

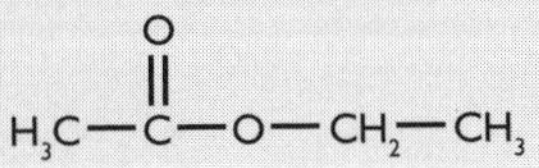

$$H_3C-C(=O)-O-CH_2-CH_3$$

Combined techniques

Analytical chemistry is like detective work. Pieces of evidence are gathered from different places. Usually no single piece of evidence is conclusive, but when all the evidence is combined, it is definitive.

Questions
&
Answers

This section contains questions similar in style to those you can expect to see in your Unit F324 examination. The limited number of questions means that it is impossible to cover all the topics and all the question styles, but they should give you a flavour of what to expect. The responses that are shown are real students' answers to the questions. Candidate A is an A/B-grade student and Candidate B is a B/C-grade student.

There are several ways of using this section. You could:

- hide the answers to each question and try the question yourself. It needn't be a memory test — use your notes to see if you can make all the points that you ought to make
- check your answers against the candidates' responses and estimate the likely standard of your response to each question
- check your answers against the examiner's comments to see if you can appreciate where you might have lost marks
- take on the role of the examiner and mark each of the responses yourself and then check to see if you agree with the marks awarded by the examiner

Examiner comments

All candidate responses are followed by examiner's comments. These are preceded by the symbol 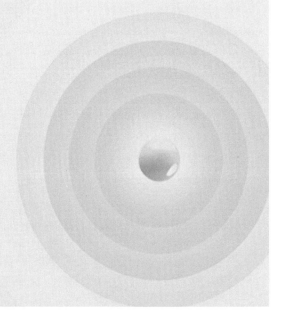 and indicate where credit is due. In the weaker answers, they also point out areas for improvement, specific problems and common errors, such as lack of clarity, irrelevance, misinterpretation of the question and mistaken meanings of terms.

Polymers

Time allocation: 7–8 minutes

Propene is an important industrial chemical essential for the production of a wide range of polymers, plastics and fibres. By far the greatest use of propene is as the monomer polymerised to poly(propene).

(a) (i) **Draw the monomer propene.** (1 mark)

    (ii) **Draw a section of poly(propene) to show *two* repeat units.** (2 marks)

(b) There are difficulties caused by waste polymers such as poly(propene). Not only are they non-biodegradable, but when burned they produce a wide range of toxic fumes, such as acrolein (CH_2=CHCHO), which has a choking odour and is the major cause of death by suffocation in house fires. Identify the *two* functional groups present in acrolein and describe how you could test to show the presence of each group. Describe what you would see with each test. (6 marks)

Total: 9 marks

■ ■ ■

Candidates' answers to Question 1

Candidate A

(a) (i)

$$
\begin{array}{cc}
CH_3 & H \\
| & | \\
C & = C \\
| & | \\
H & H
\end{array}
$$

Candidate B

(a) (i) CH_3CH=CH_2

   ✎ Both candidates score 1 mark.

Candidate A

(a) (ii)

$$
\begin{array}{cccc}
CH_3 & H & CH_3 & H \\
| & | & | & | \\
C & C & C & C \\
| & | & | & | \\
H & H & H & H
\end{array}
$$

Candidate B

(a) (ii)

$$
\begin{array}{cccccc}
H & H & H & H & H & H \\
| & | & | & | & | & | \\
-C & C & C & C & C & C- \\
| & | & | & | & | & | \\
H & H & H & H & H & H
\end{array}
$$

Candidate A scores 1 mark, but loses 1 mark carelessly by not showing the 'end' bonds on the terminal carbon atoms. These are essential as they indicate that the polymer continues on either side. It is worth remembering that a carbon is always drawn to show that it has four bonds. Candidate B scores no marks because the polymer drawn is in fact three repeat units of poly(ethene) and not two repeat units of poly(propene). This is a common error.

Candidate A

(b) Alkene and aldehyde. Alkenes decolorise bromine water. Aldehydes give a silver mirror with Tollens' reagent.

Candidate B

(b) Acrolein contains a carbonyl, which will react with 2,4-dinitrophenylhydrazine to produce a red precipitate. It also contains a $C=C$, which turns bromine clear.

For this question, Candidate A scores 4 out of the 6 marks, while Candidate B scores 3 out of 6. Both candidates have identified the functional groups and have stated a suitable reagent, but neither has stated how the test would be carried out. For instance, the bromine test only works if it is added dropwise to avoid being in excess. Candidate B loses another mark by using the word 'clear' in the test for the alkene. Bromine is a clear solution; it just happens to be a clear *brown* solution — since it is already clear, it cannot 'turn clear'.

Throughout the rest of this book, it is important to analyse where each candidate went wrong and to identify what needs to be done to pick up the extra marks. In question 1, Candidate A scores 6 out of 9 marks, but Candidate B gains only 4 out of 9 marks. Candidate B's overall response to this question is equivalent to a grade D, but, with a little care, the score could have been increased by 1 or 2 marks to the B/C borderline.

Question 2

Nitration of arenes

Time allocation: 11–12 minutes

Methylbenzene is an important industrial chemical. It is used in the production of polyurethane plastic foams or fibres such as *lycra*. The production of such foams and fibres involves the nitration of methylbenzene.

(a) Methylbenzene undergoes electrophilic substitution with the nitronium ion (NO_2^+) to form 4-nitromethylbenzene $(CH_3C_6H_4NO_2)$.
  (i)   With the aid of curly arrows, show the mechanism for the formation of 4-nitromethylbenzene. (3 marks)
  (ii) In a laboratory preparation, 9.20 g of methylbenzene were used and 5.48 g of pure 4-nitromethylbenzene were isolated. Calculate the percentage yield and the atom economy of the reaction. (5 marks)

(b)

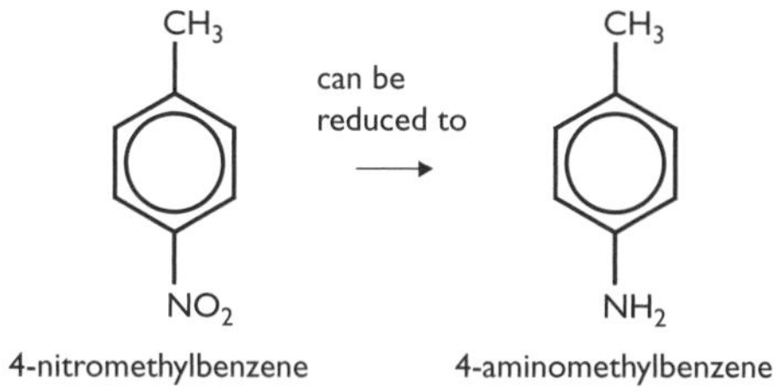

CH$_3$

can be reduced to

CH$_3$

NO$_2$

NH$_2$

4-nitromethylbenzene   4-aminomethylbenzene

  (i)   Suggest a suitable reducing agent or a suitable reducing mixture for this reaction. (1 mark)
  (ii) Construct a balanced equation for this reduction. Use [H] to represent the reducing agent. (2 marks)

(c) There are six structural isomers of dinitromethylbenzene $(CH_3C_6H_3(NO_2)_2)$. Four are drawn for you; draw the structures of the other *two* isomers. (2 marks)

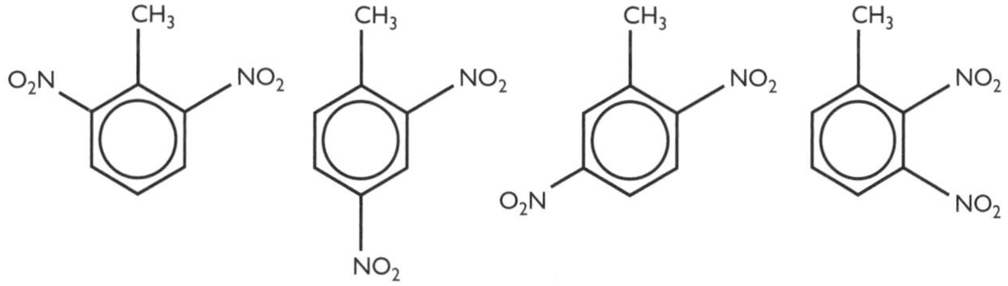

CH$_3$ CH$_3$ CH$_3$ CH$_3$

O$_2$N   NO$_2$ NO$_2$ NO$_2$ NO$_2$

O$_2$N   NO$_2$

NO$_2$

(d) The manufacture of *lycra* involves one of these six isomers. A small section of *lycra* is shown opposite:

Draw the structure of the isomer of dinitromethylbenzene used in the manufacture of *lycra*. (1 mark)

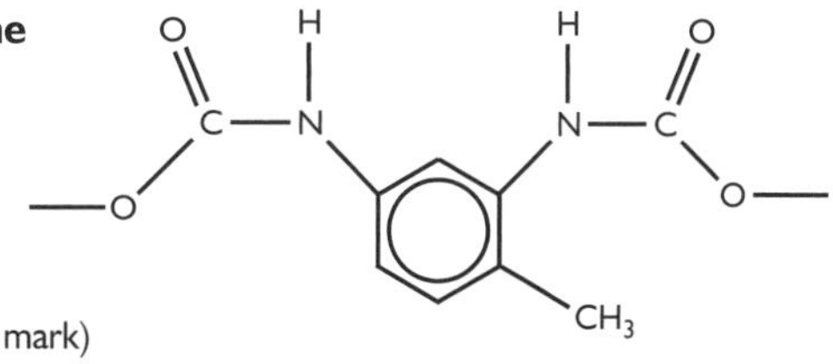

O   H   H   O
‖ | | ‖
C—N   N—C
| |
—O   O—

CH$_3$

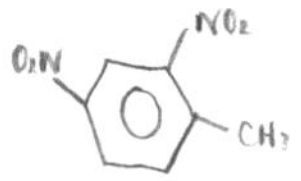

O$_2$N   NO$_2$

CH$_3$

Total: 14 marks

2
question

Candidates' answers to Question 2

Candidate A

(a) (i)

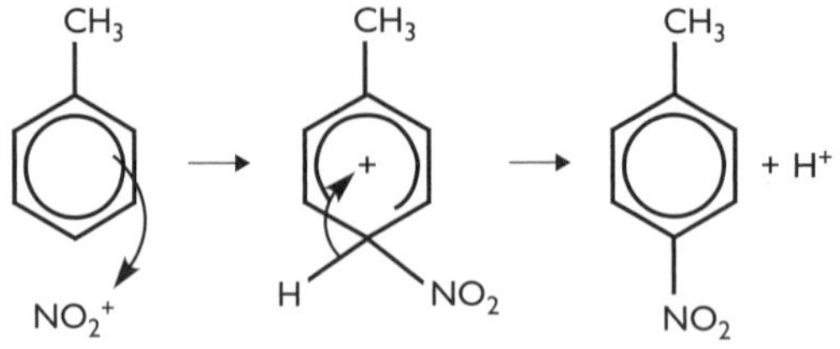

Candidate B

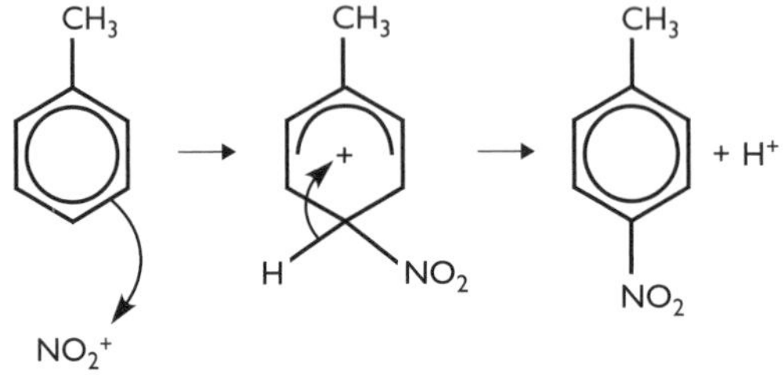

> Candidate A gains all 3 marks, but Candidate B scores only 1 mark. The first marking point is the curly arrow from the π delocalised ring to the NO_2^+ ion. Candidate B's curly arrow doesn't start at the π delocalised ring. The second marking point is the intermediate, which must show clearly the net positive charge and the breaking of the π delocalised ring at the carbon being attacked. Candidate B has carelessly drawn the broken π delocalised ring over three carbon atoms. The final marking point is for the curly arrow showing the reforming of the π delocalised ring. Both candidates score the final mark.

Candidate A

(a) (ii) Moles of reagent = 9.20/93 = 0.099

Moles of product = 5.48/139 = 0.039

Percentage yield = (0.039/0.099) × 100 = 39.4%

Atom economy = (mass of products/mass of reactants) × 100
= (139/93) × 100 = 149.5%

Candidate B

(a) (ii) Percentage yield = 40%

Atom economy = 67%

> Candidate A scores 1 out of 3 marks for the percentage yield calculation. The relative molecular masses of both chemicals are incorrect, so Candidate A loses 2 marks. However, credit is given for the rest of the calculation. The equation used for atom economy is incorrect and a percentage exceeding 100% should have alerted Candidate A that an error had been made. However, Candidate A has displayed good examination technique and has shown all the working. This

enables the examiner to see where mistakes were made and to award marks for the correct processing of the numbers. Candidate B gains all 3 marks for the correct percentage yield, but nothing for the atom economy. However, Candidate B shows poor examination technique by not showing the working. Neither scores any marks for the atom economy. It is worth remembering that all A2 papers are synoptic and may test knowledge and understanding from the AS specification.

Atom economy = (mass of desired product/mass of all products) × 100. The products are $CH_3C_6H_4NO_2$ and H_2O, so the atom economy: (137/155) × 100 = **88.4%**

Candidate A

(b) (i) Tin and concentrated HCl

Candidate B

(b) (i) $LiAlH_4$

> Both candidates gain the mark. Candidate A has quoted the reducing agent mentioned in the specification, but $LiAlH_4$ would also work and so earns the mark.

Candidate A

(b) (ii)

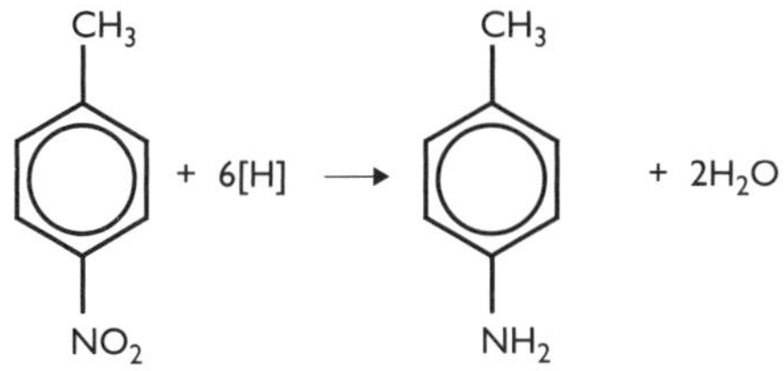

Candidate B

(b) (ii)

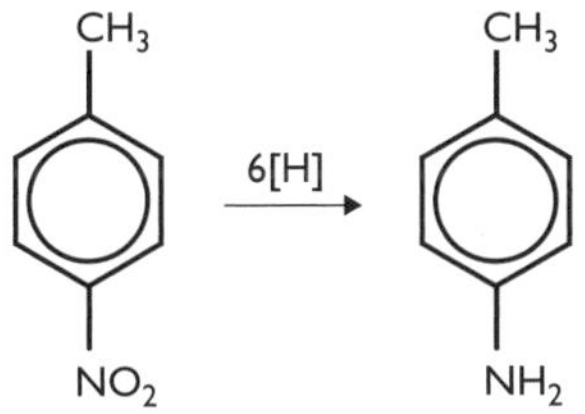

> Candidate A scores both marks. Candidate B loses 1 mark by forgetting to include water in the equation, which is, therefore, not balanced.

Candidate A

(c)

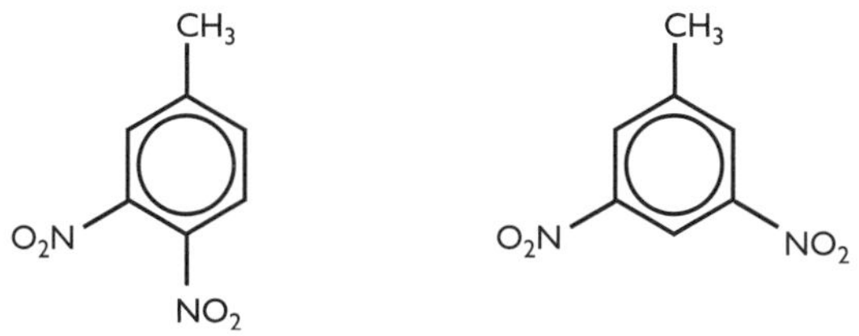

2

question

Candidate B

(c)

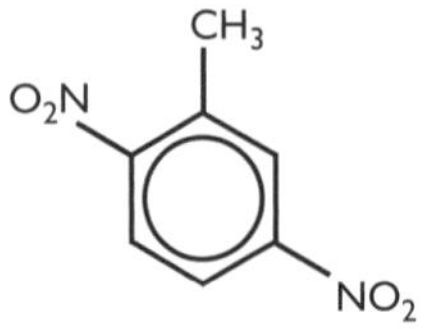

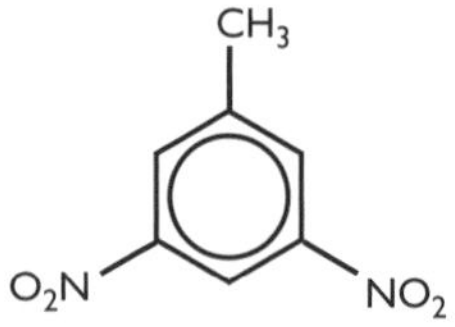

📝 Candidate A gains both marks. Candidate B scores 1 mark only, because the first structure drawn is the same as the third structure given in the question.

Candidate A

(d)

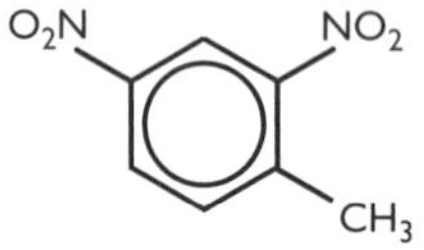

Candidate B

(d)   2,4-dinitromethylbenzene

📝 Candidate A shows better examination technique by copying sensibly the structure given in the question and replacing the amide groups with the NO_2 group. Candidate B works out the correct compound but loses the mark by naming rather than drawing it.

📝 **Candidate B seems to be a bright candidate, but is perhaps trying to be too clever and cut corners. This is evident in (a)(ii), when no working is shown. Candidate B scored all 3 marks for the percentage yield calculation, but could have easily lost them all. In the final part, Candidate B does more than is necessary by naming the product. This means that the candidate had first to deduce the structure and then name it, apparently in his/her head, since there was no evidence of working. Candidate A shows understanding and carefully uses all the information in the question. The net result is that Candidate A scores 10 out of 14 marks, a grade-B answer, while Candidate B gains 7, which is borderline D/C-grade standard.**

Carbonyls and carboxylic acids

Time allocation: 9–10 minutes

Propan-1-ol can be oxidised both to propanal and to propanoic acid.

(a) (i) State a suitable oxidising mixture. (2 marks)
(ii) State what you would see during the oxidations. (1 mark)
(iii) Using [O] to represent the oxidising mixture, write a balanced equation to show the oxidation of propan-1-ol to propanal. (1 mark)
x (iv) Similarly, write a balanced equation for the oxidation of propanal to propanoic acid. (1 mark)

(b) Describe a simple chemical test that would distinguish between propanal and propanoic acid. State what you would see. (2 marks)

(c) Compound **X** contains C, H and O only. Its relative molecular mass is 102. When 5.1 g of compound **X** is burnt, in excess oxygen, 6.0 dm^3 of CO_2 are produced. Compound **X** can be hydrolysed to form propan-1-ol and one other organic compound. Use all of the information in the question to deduce the molecular formula of compound **X**. Draw the structure of compound **X**. Show all of your working. (5 marks)

Total: 12 marks

■ ■ ■

Candidates' answers to Question 3

Candidate A

(a) (i) Acidified potassium dicromate

Candidate B

(a) (i) $H^+/Cr_2O_7^-$

☑ Candidate A gains both marks, 1 for the acid and 1 for the dichromate, even though 'dichromate' isn't spelt correctly. Candidate B scores the mark for the acid (H^+), but loses the mark for the dichromate because the charge on the ion should be 2^-, not 1^-. Spelling is only penalised when marks are allocated for quality of written communication, but incorrect formulae will always be penalised.

Candidate A

(a) (ii) Orange to green

Candidate B

(a) (ii) Turns green

☑ Candidate A gains the mark, but Candidate B has only stated half of what would be observed, and thus scores no marks.

3
question

Candidate A

(a) (iii)

$$CH_3CH_2CH_2OH + [O] \rightarrow H-C-C-C$$

Candidate B

(a) (iii)

$$CH_3CH_2CH_2OH \xrightarrow{[O]} CH_3CH_2CHO$$

☑ Both candidates score no marks, because neither of them has balanced the equation. This is a common error. Both have correctly identified the organic product but forgotten to include water as a product.

Candidate A

(a) (iv)

Candidate B

(a) (iv)

$$CH_3CH_2CHO \xrightarrow{[O]} CH_3CH_2COOH$$

☑ Both candidates score the mark because the equation is automatically balanced.

Candidate A

(b) Add a solution of sodium hydrogencarbonate to both. The propanoic acid will react and you will see bubbles of CO_2. The propanal doesn't react.

Candidate B

(b) The pH of the acid will be lower than the pH of propanal.

☑ Candidate A gains both marks by describing a suitable chemical test and appropriate observations. Candidate B has described how pH could be used to distinguish between the two chemicals. The pH is most easily measured using a pH meter, which is not a *chemical* test. Candidate B would probably be awarded 1 of the 2 marks, but may fail to score.

Candidate A

(c) Moles of X = 5.1/102 = 0.05

Moles of CO_2 = 6/24 = 0.25

1 mole of X produces 5 moles of CO_2, therefore X must have five carbons.

102 − 60 = 42; hence X contains two oxygens and 10 hydrogens.

Thus, the molecular formula is $C_5H_{10}O_2$.

X is hydrolysed to form an alcohol; hence it is an ester.

Candidate B

(c) Compound X is propyl ethanoate, because it hydrolyses to form propan-1-ol and ethanoic acid.

$$CH_3COOCH_2CH_2CH_3 + H_2O \rightarrow CH_3COOH + CH_3CH_2CH_2OH$$

📝 The marking points are: 1 mark for moles of X and CO_2 ✓; 1 mark for giving the ratio of moles of X and CO_2 as 1:5 ✓; 1 mark for correct molecular formula, $C_5H_{10}O_2$ ✓; and 1 mark for deducing that X is an ester ✓; 1 mark for identifying propyl ethanoate ✓. Candidate A has been methodical and has followed the guidelines given in the question. He/she scores the first 4 marking points, but then carelessly draws ethyl propanoate instead of propyl ethanoate. Hence, Candidate A scores 4 out of 5 marks. Candidate B is clearly able but has not followed the instructions in the question. The only marks that can be awarded are the final 3 marking points. Candidate B scores 3 out of 5 marks. He/she has not used the information about the mass of X used and the volume of CO_2 produced. Note that examiners never give information unless it is relevant to the question.

📝 **An examiner may feel sympathetic for Candidate B. Clearly, he/she has a good understanding of the work. However, poor examination technique results in consistent under-achievement. This is particularly evident in part (c) where Candidate B quickly sees the answer and correctly deduces the identity of compound X, but doesn't follow the instructions in the question to use *all* of the information given. Consequently, Candidate B scores only 6 out of a possible 12, which equates to a grade D. By contrast, Candidate A is much more methodical and scores 10 out of 12 marks. The net result is that Candidate A scores grade A on this question, while Candidate B achieves only a grade D.**

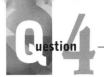

Amino acids

Time allocation: 6–7 minutes

This question is about the amino acids in the table below.

Key	Name	Structure
Ala	Alanine	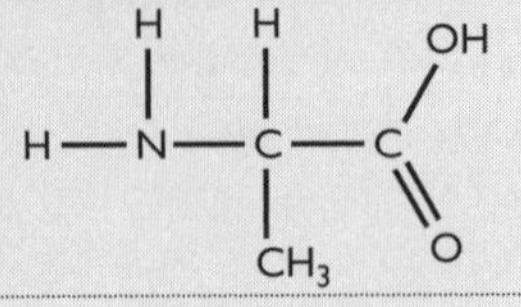
Gly	Glycine	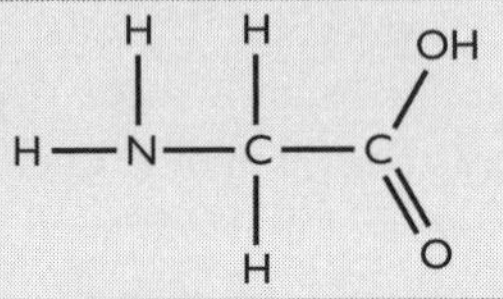
Phe	Phenylalanine	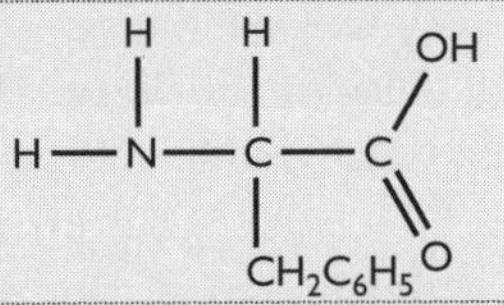
Cys	Cysteine	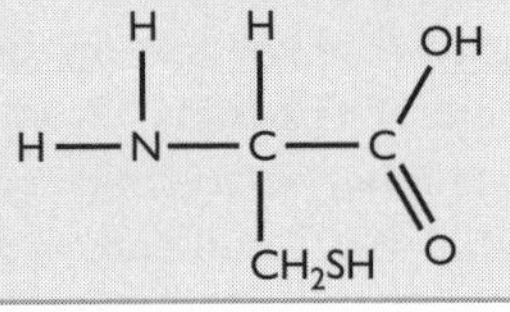

(a) A dipeptide is hydrolysed into its amino acids and the amino acids are identified by chromatography. The results are shown on the chromatogram below.

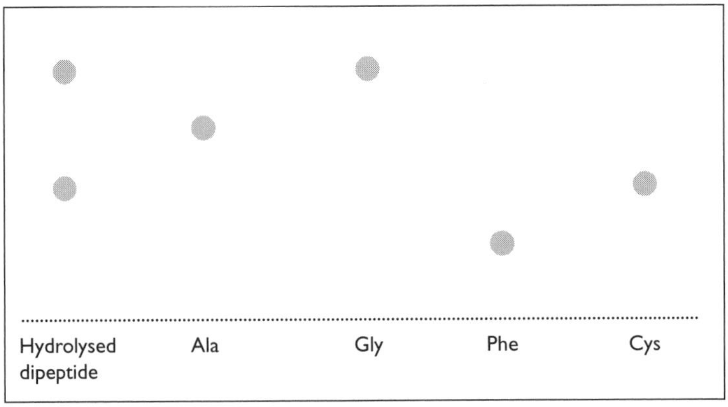

(i) Identify the two amino acids present in the dipeptide. (1 mark)

(ii) Suggest two possible structures of the dipeptide. (2 marks)

(b) The pHs at which the zwitterions of glycine, alanine and phenylalanine exist are:

- glycine, 6.0
- alanine, 6.0
- phenylalanine, 5.5

  (i) Draw the ions formed by glycine at pH = 6.0, alanine at pH = 10.0 and phenylalanine at pH = 2.0. (3 marks)

  (ii) Suggest why a solution of glycine does *not* conduct electricity at pH = 6.0. (1 mark)

  (iii) Suggest why a solution of alanine does conduct electricity at pH = 10.0. *... negative ∴ → ANODE* (1 mark)

  Total: 8 marks

■ ■ ■

Candidates' answers to Question 4

Candidate A

(a) (i)   Glycine and cysteine

Candidate B

(a) (i)   Gly and Cys

⟨e⟩ Both candidates gain the mark.

Candidate A

(a) (ii)

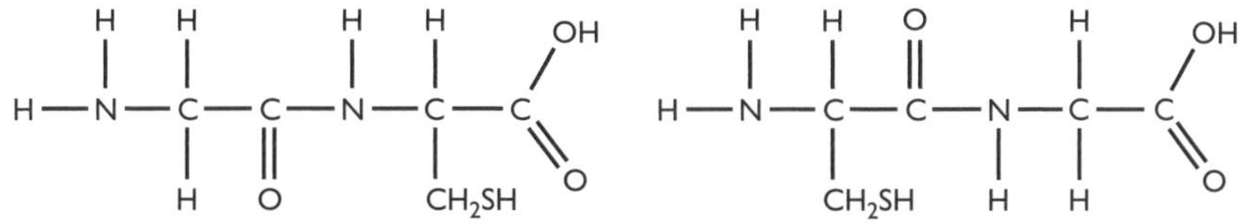

Candidate B

(a) (ii)

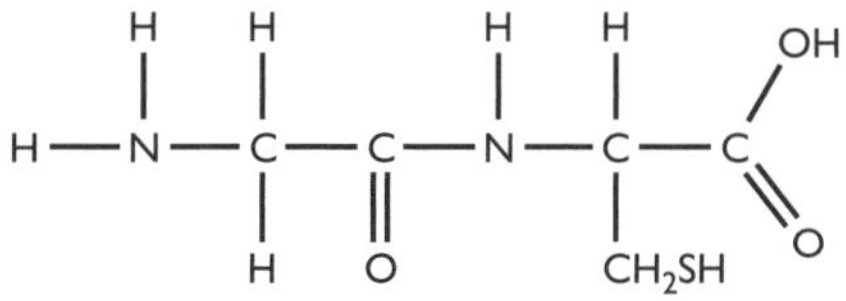

⟨e⟩ Candidate A gains both marks, but Candidate B has drawn only one of the two peptides, for 1 mark. Candidate B needs to slow down and to read the question carefully. There are 2 marks, which is a clear indication that two responses are required.

question

Candidate A
(b) (i)

pH = 6.0

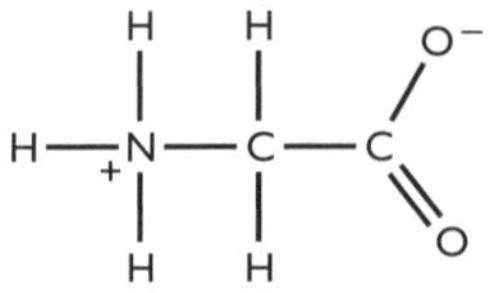

pH = 10.0

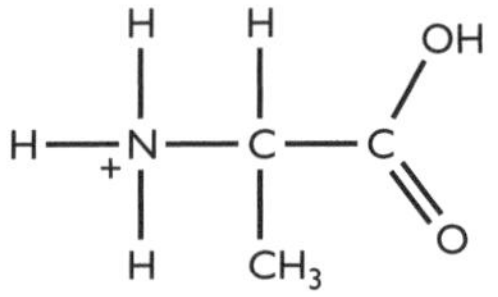

pH = 2.0

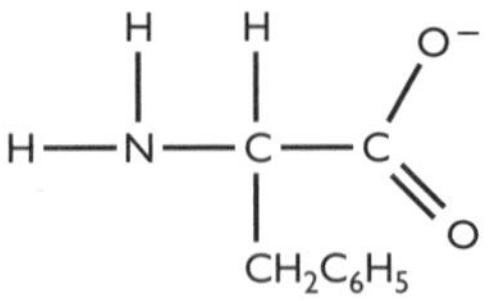

Candidate B
(b) (i)

pH = 6.0

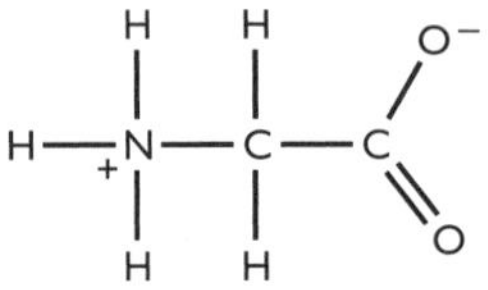

pH = 10.0

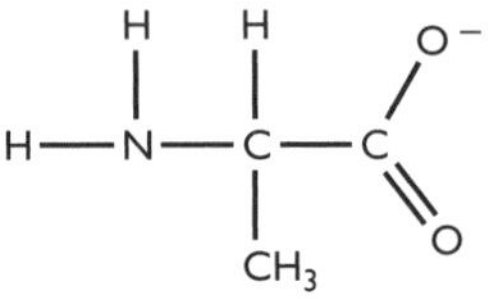

pH = 2.0

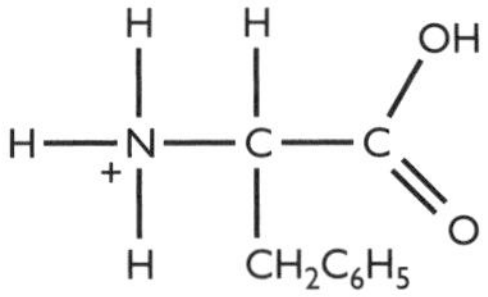

> Candidate A scores 1 mark for the zwitterion at pH = 6.0. However, the charges on the other two ions are incorrect. Candidate B gains all 3 marks.

Candidate A

(b) (ii) The zwitterion has no net charge.

Candidate B

(b) (ii) It is a poor conductor.

> Candidate A gains the mark. Candidate B has simply restated the question and is awarded no marks.

Candidate A

(b) (iii) The ion is positive and will therefore move to the cathode.

Candidate B

(b) (iii) It is a good conductor.

> Candidate A gains the mark. Candidate B has made the same mistake as in part **(b)(ii)** and scores nothing.

The most difficult part of this question is (b)(i). Candidate A has mixed up the ions, but Candidate B has scored full marks. However, overall, Candidate A gains 6 marks out of 8, while Candidate B scores 5 marks.

Organic synthesis

Time allocation: 7–8 minutes

Consider the reaction scheme shown below:

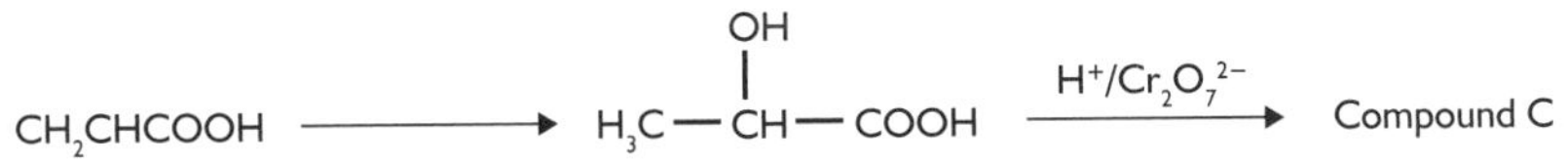

$$CH_2CHCOOH \longrightarrow H_3C-\overset{\overset{\displaystyle OH}{|}}{CH}-COOH \xrightarrow{H^+/Cr_2O_7{}^{2-}} \text{Compound C}$$

Compound A   Compound B

(a) (i) Name compound A.   (1 mark)
   (ii) Write a balanced equation for the conversion of compound A into
         compound B.   (1 mark)

(b) Compound B, lactic acid, can be found in cheese and can exist as one of
    two stereoisomers.
   (i) Draw the two stereoisomers of compound B.   (2 marks)
   (ii) Explain whether or not compound B, prepared by the reaction
        scheme, would contain both stereoisomers or not.   (1 mark)

(c) (i) Draw the structure of compound C.   (1 mark)
   (ii) Compound C can be reduced to propane-1,2-diol. Using [H] to
         represent the reducing agent, construct a balanced equation for
         this reduction.   (3 marks)

Total: 9 marks

■ ■ ■

Candidates' answers to Question 5

Candidate A
(a) (i) Propenoic acid

Candidate B
(a) (i) Propanoic acid

e Neither candidate scores the mark. At first glance this looks easy, but it is not.
     Candidate B does not notice that there has to be a double bond between the CH_2
     and the CH in the $CH_2CHCOOH$. Candidate A's attempt is better, but still not
     correct. Carboxylic acids always end in '-oic acid' and the —COOH group always
     occurs at carbon atom 1, meaning that the C=C double bond must be between
     the second and third carbons. Hence, the correct name is prop-2-enoic acid.

Candidate A
(a) (ii)

$$CH_2CHCOOH \quad + \quad H_2O \longrightarrow H_3C-\overset{\overset{\displaystyle OH}{|}}{CH}-COOH$$

Compound A   Compound B

Candidate B

(b) (i) $C_3H_4O_2 + H_2O \rightarrow C_3H_6O_3$

> Both candidates gain 2 marks. Candidate A has shown good exam technique and copied the information in the question. Candidate B has made life hard by using molecular formulae that are not required. Many candidates do this and they often make mistakes. Use the information in the question, and whenever possible avoid mistakes by copying the formulae/structures from the question.

Candidate A

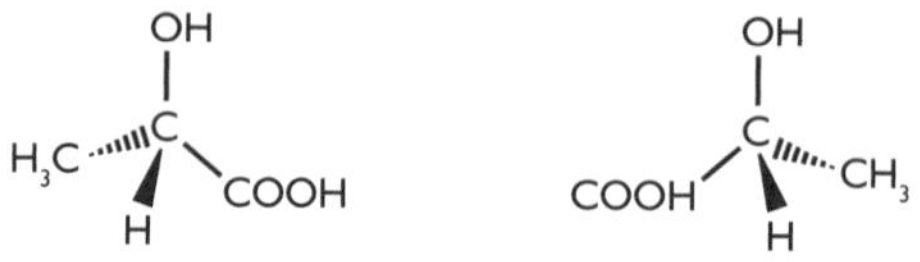

Candidate B

> Both score 1 mark, but for different reasons. Candidate A has drawn a correct three-dimensional structure, but loses a mark for the structure on the right because the carboxylic acid group appears to be bonded to the central carbon via the H. This is known as an incorrect bond linkage. Candidate B also gains only 1 mark. The four groups are correctly bonded and the two isomers are mirror images, but the three-dimensional arrangement is incorrect.

>   ———— indicates a bond in the plane of the paper
>
>   ◄■■■ indicates a bond in front of the plane of the paper 
>
> ||||···· indicates a bond behind the plane of the paper 

> If two bonds are drawn as '——', then they must be adjacent to each other. This is difficult to visualise, but easy to see if you build a model of the two stereoisomers.

Candidate A

(b) (ii)   It forms a mixture of both, because the addition across the double bond could be from either side; it isn't stereo-specific. You only get one isomer if enzymes are used.

Candidate B

(b) (ii)   An equal amount of each isomer would be formed, and therefore they would cancel each other out.

> This is a difficult concept and Candidate A has given a good answer. Candidate B has shown good understanding, but does not score the mark. This is because there is no explanation of why a mixture is obtained, simply a statement that it would be a 50:50 mixture.

Candidate A

(c) (i)

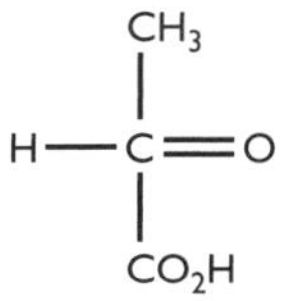

$$\begin{array}{c} CH_3 \\ | \\ H-C=O \\ | \\ CO_2H \end{array}$$

Candidate B

(c) (i)

$$\begin{array}{c} CH_3 \\ | \\ C=O \\ | \\ C=O \\ | \\ H \end{array}$$

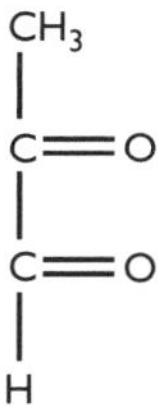

Neither candidate scores the mark. It is important to recognise that the functional group attached to the middle carbon atom is a secondary alcohol and is therefore oxidised to a ketone. The other functional group is a carboxylic acid, which is not oxidised. Both candidates have recognised that the alcohol will be oxidised. However, Candidate A loses the mark because the central carbon atom contains five bonds. Candidate B loses the mark because the carboxylic acid has been reduced back to an aldehyde. The correct structure is:

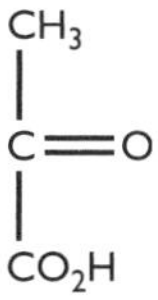

$$\begin{array}{c} CH_3 \\ | \\ C=O \\ | \\ CO_2H \end{array}$$

Candidate A

(c) (ii)

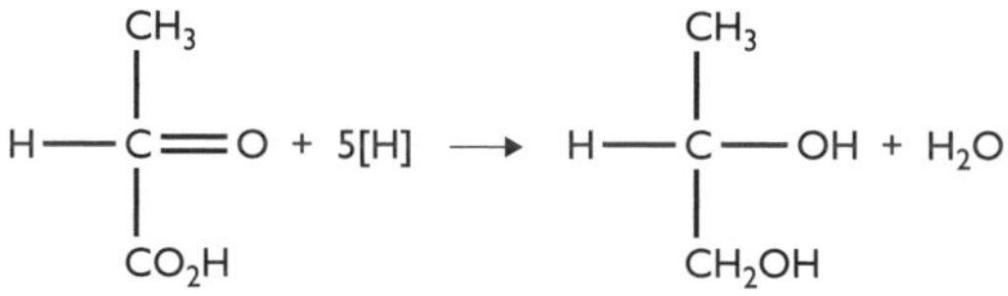

$$\begin{array}{c} CH_3 \\ | \\ H-C=O \\ | \\ CO_2H \end{array} + 5[H] \longrightarrow \begin{array}{c} CH_3 \\ | \\ H-C-OH \\ | \\ CH_2OH \end{array} + H_2O$$

Candidate B

(c) (ii)

$$\begin{array}{c} CH_3 \\ | \\ C=O \\ | \\ C=O \\ | \\ H \end{array} + 4[H] \longrightarrow \begin{array}{c} CH_3 \\ | \\ H-C-OH \\ | \\ CH_2OH \end{array}$$

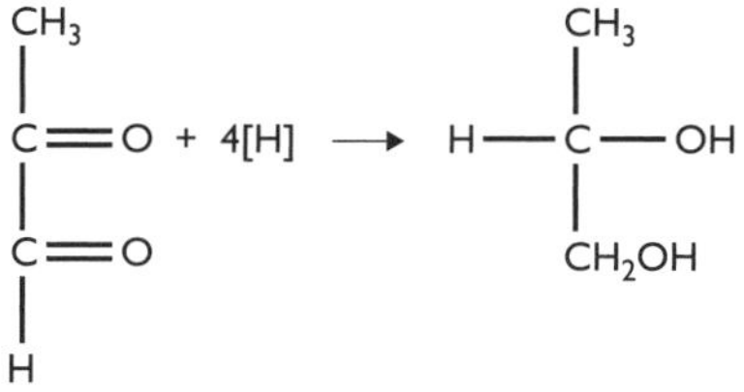

The three marking points are: correct formula for propane-1,2-diol ✓; H_2O as a product ✓; and the equation correctly balanced ✓. Applying this mark scheme, Candidate A scores 2 marks, while Candidate B gains 1 mark.

This is a difficult question, and Candidate A scores 6 out of 9 marks, while Candidate B scores 4. Questions like this are often asked. Examiners are able to use unfamiliar molecules to test routine functional-group chemistry. The key to successfully answering questions like these is to identify the essential functional groups and to concentrate on the chemistry of the group and ignore the rest of the molecule.

Carboxylic acids, esters and aldehydes

Time allocation: 9–10 minutes

Compound A has the structure shown below:

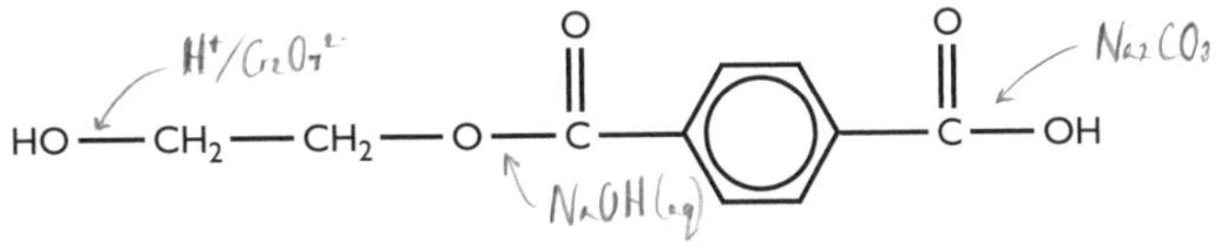

$$HO \xleftarrow{H^+/Cr_2O_7^{2-}} CH_2 - CH_2 - O \underset{\underset{NaOH(aq)}{\uparrow}}{\overset{\overset{O}{\|}}{-C}} - \bigcirc - \overset{\overset{O}{\|}}{C} - OH \xleftarrow{Na_2CO_3}$$

(a) Deduce the empirical formula and molecular formula of compound A. *(2 marks)*

(b) Suggest three different reagents that will react with compound A. Identify the organic product(s) of each reaction and name the type of reaction involved. *(9 marks)*

Total: 11 marks

■ ■ ■

Candidates' answers to Question 6

Candidate A

(a) The molecular formula is $C_{10}H_{10}O_5$ and the empirical formula is C_2H_2O.

Candidate B

(a) Empirical formula $= C_2H_2O$ and molecular formula $= HOCH_2CH_2O_2CC_6H_4CO_2H$

> ☑ Candidate A gains both marks, but Candidate B only scores 1 out of 2. Molecular formulae should always be written in the form $C_xH_yO_z$, so that the elements that make up the compound are all grouped together.

Candidate A

(b) It is oxidised with $H^+(aq)/Cr_2O_7^{2-}(aq)$ to give:

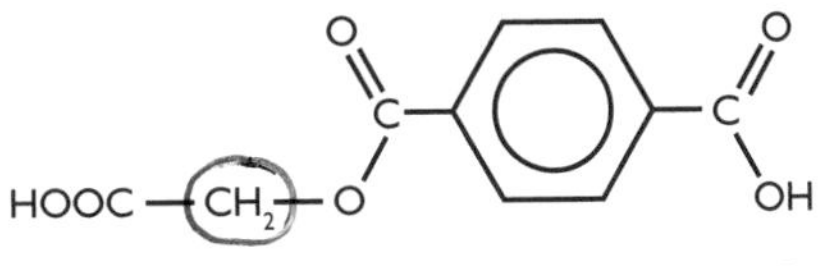

$$HOOC - \left(CH_2\right) - O \underset{}{\overset{\overset{O}{\|}}{C}} - \bigcirc - \overset{\overset{O}{\|}}{C} - OH$$

It is neutralised by NaOH(aq) to give:

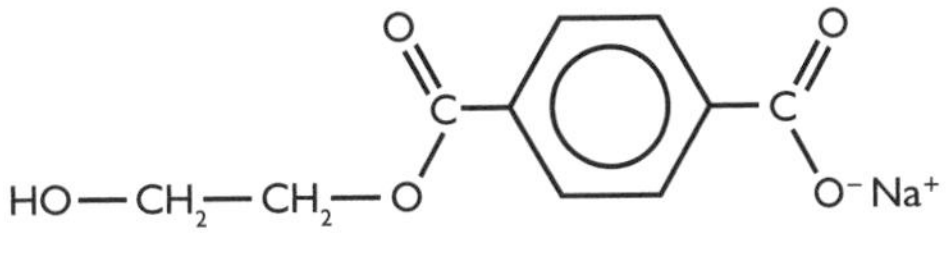

$$HO - CH_2 - CH_2 - O \underset{}{\overset{\overset{O}{\|}}{C}} - \bigcirc - \overset{\overset{O}{\|}}{C} - O^-Na^+$$

It is hydrolysed by $H_2SO_4(aq)$ to give:

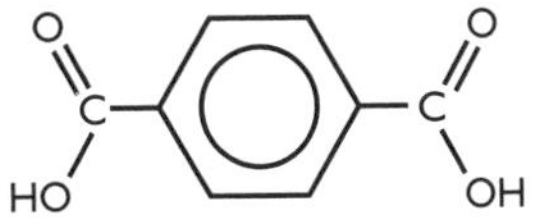

$$HO \overset{\overset{O}{\|}}{\underset{}{C}} - \bigcirc - \overset{\overset{O}{\|}}{\underset{}{C}} OH$$

Candidate B

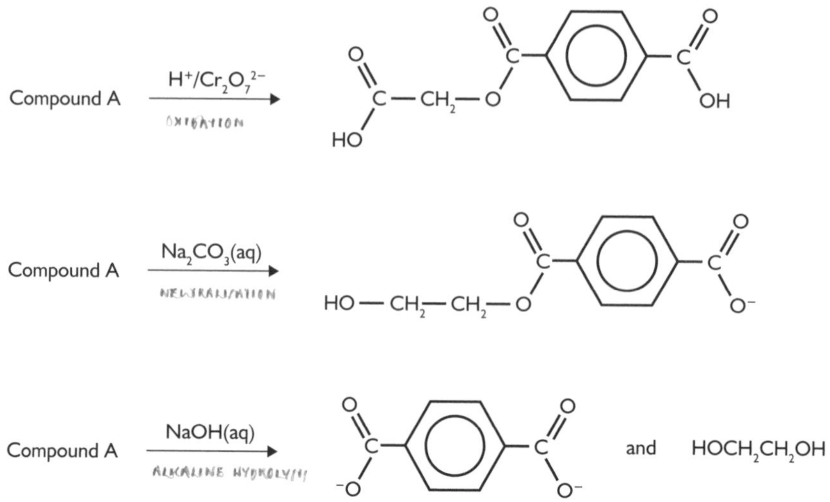

Compound A $\xrightarrow[\text{OXIDATION}]{H^+/Cr_2O_7^{2-}}$

Compound A $\xrightarrow[\text{NEUTRALISATION}]{Na_2CO_3(aq)}$

Compound A $\xrightarrow[\text{ALKALINE HYDROLYSIS}]{NaOH(aq)}$ and $HOCH_2CH_2OH$

Candidate A scores a total of 8 marks. Candidate B is awarded 6 marks, but appears to be the better chemist. The two candidates have adopted different approaches. Candidate A carefully follows the instructions in the question, and for each of the three reactions gives the reagent, organic product and type of reaction. In the first reaction, all three are correct, so 3 marks are awarded. In the second reaction, NaOH will neutralise the carboxylic acid, but it will also hydrolyse the ester. However, this answer still scores 2 marks. Candidate A drops 1 mark in the last reaction by showing only one of the organic products. Candidate B shows excellent knowledge of chemistry, but loses a mark in each reaction by not stating the type of reaction involved. Failure to carefully read the question has cost Candidate B one-third of the marks.

This is a difficult question making use of functional-group chemistry in an unfamiliar situation. Candidate A has done well and scored 9 out of 11 marks. Candidate B has scored 7 out of 11, but part (b) shows clearly that Candidate B is the better chemist and with more care could have scored all 11 marks. Every mark is important — 9 out of 11 marks means grade A for Candidate A, but Candidate B scores just over 60% of the marks, which is a grade C.

Amides, esters and chirality

Time allocation: 11–12 minutes

(a) Aspartame, shown below, can be used as an artificial sweetener.

H_2N — $\overset{*}{C}H$ — $\overset{O}{\underset{\|}{C}}$ — N — $\overset{*}{C}H$ — $\overset{O}{\underset{\|}{C}}$ — O — CH_3

AMINE CH_2 PEPTIDE BOND H CH_2 ESTER

CO_2H

CARBOXYLIC ACID

(benzene ring)

(i)   Aspartame contains five functional groups, including the benzene ring. Name the other *four* functional groups. (4 marks)

(ii) *Two* of the four functional groups can be hydrolysed. Circle these groups on the diagram above. (2 marks)

(iii) Show the structures of the organic products formed by the acid hydrolysis of aspartame. (3 marks)

(b) (i) Aspartame has two chiral carbon atoms. Identify each with an asterisk (*). (2 marks)

(ii) Explain what is meant by the term *chiral* and deduce the number of possible stereoisomers. (2 marks)

(c) Aspartame can be made from aspartic acid (shown below):

H_2N — CH — $\overset{O}{\underset{\|}{C}}$ — OH

CH_2

CO_2H

Suggest the structure of a compound that could react with aspartic acid to make aspartame. (1 mark)

Total: 14 marks

■ ■ ■

Candidates' answers to Question 7

Candidate A

(a) (i)   Amide, carboxylic acid, ester, peptide

Candidate B

(a) (i)   Amine, amide, ketone, carboxylic acid, ester

7

question

📝 Candidate A gains 3 of the 4 marks. One mark is lost because amide and peptide are the same. Many candidates adopt the technique used by Candidate B. The question asks for *four* functional groups, so candidates think that they are hedging their bets by listing *five* functional groups. There are *only four* functional groups. By writing five, Candidate B has automatically lost 1 mark. The wrong answer will be marked first — the examiner will not select the correct answers from a list of alternatives.

Candidate A

(a) (ii)

$$H_2N-CH-\overset{\overset{\displaystyle O}{\|}}{C}-N-CH-\overset{\overset{\displaystyle O}{\|}}{C}-O-CH_3$$

with CH_2 and CO_2H below the first CH, H below the N, and CH_2 with phenyl ring below the second CH

Candidate B

(a) (ii)

$$H_2N-CH-\overset{\overset{\displaystyle O}{\|}}{C}-N-CH-\overset{\overset{\displaystyle O}{\|}}{C}-O-CH_3$$

with CH_2 and CO_2H below the first CH, H below the N, and CH_2 with phenyl ring below the second CH

📝 Candidate A gains both marks. Candidate B loses a mark by extending the circle to include both the amide and the amine group.

Candidate A

(a) (iii)

$$HO-\overset{\overset{\displaystyle O}{\|}}{C}-N-CH-\overset{\overset{\displaystyle O}{\|}}{C}-O-OH$$

with H below the N, and CH_2 with phenyl ring below the CH

$$H_2N-CH-OH$$

with CH_2 and CO_2H below the CH

$HOCH_3$

Candidate B

(a) (iii)

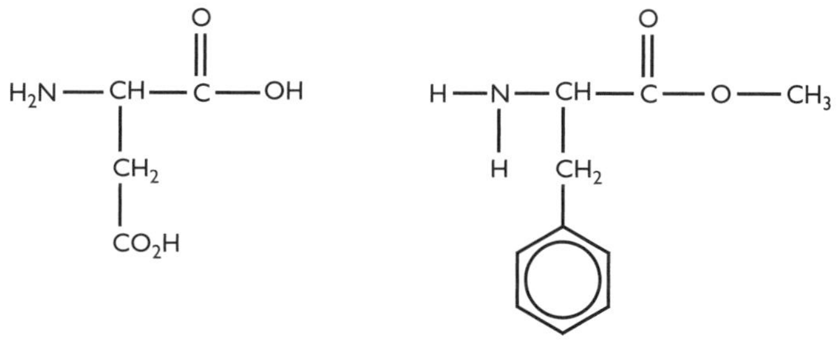

⟲ There are 3 marks allocated here, indicating clearly that there are three products. Candidate A has used this information, but Candidate B displays poor examination technique by drawing only two products. Hydrolysis is a difficult concept. Both the amide and the ester undergo hydrolysis. The bonds that break and the correct products are shown in the diagram below. However, the acid used as a catalyst for the hydrolysis will then react with NH_2 groups to form $^+NH_3$ salts. Use the answer below to work out where each candidate went wrong.

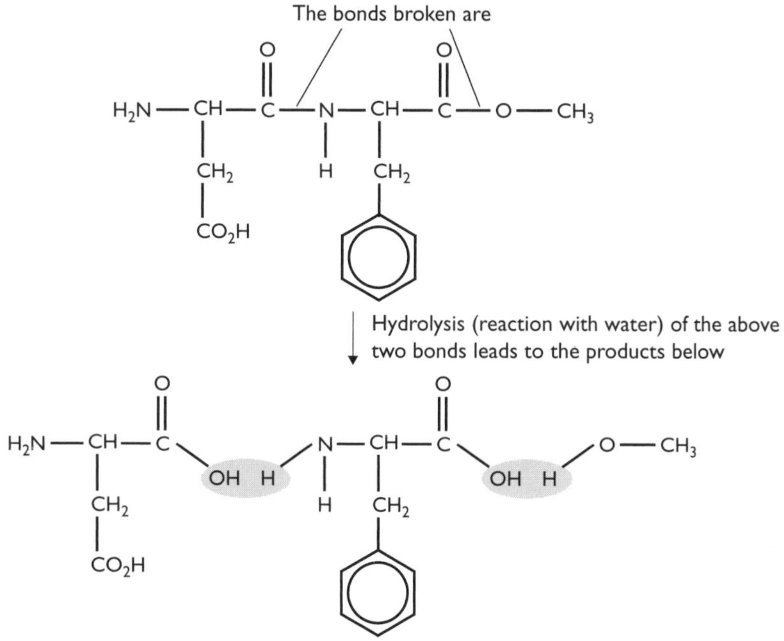

The bonds broken are

Hydrolysis (reaction with water) of the above two bonds leads to the products below

The final products using acid hydrolysis are:

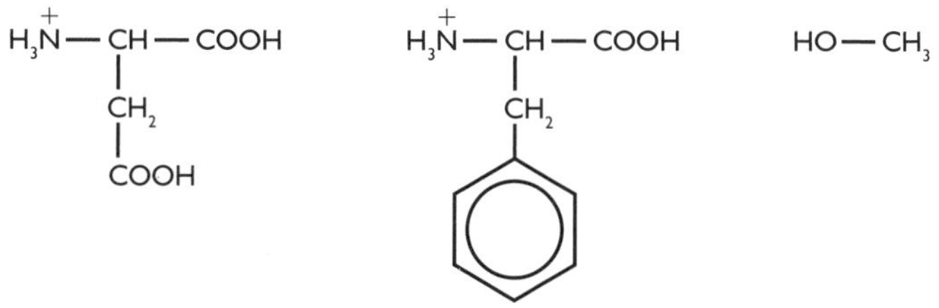

⟲ Candidate A scores just 1 mark, and Candidate B earns no marks.

question

Candidate A

(b) (i)

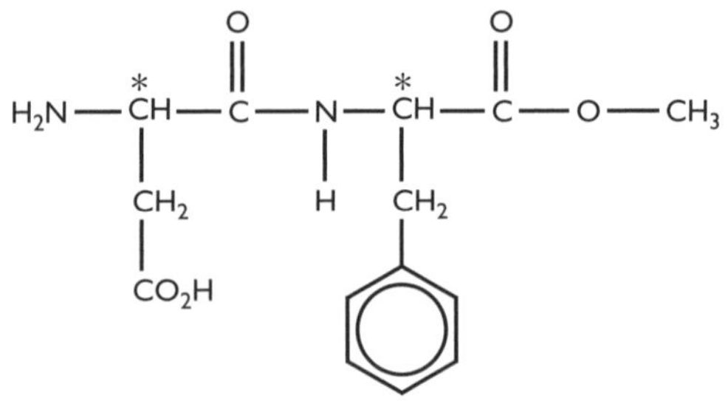

H$_2$N — CH — C(=O) — N — CH — C(=O) — O — CH$_3$

with CH$_2$—CO$_2$H branch and H, CH$_2$—phenyl branch; both CH carbons marked with *

Candidate B

(b) (i)

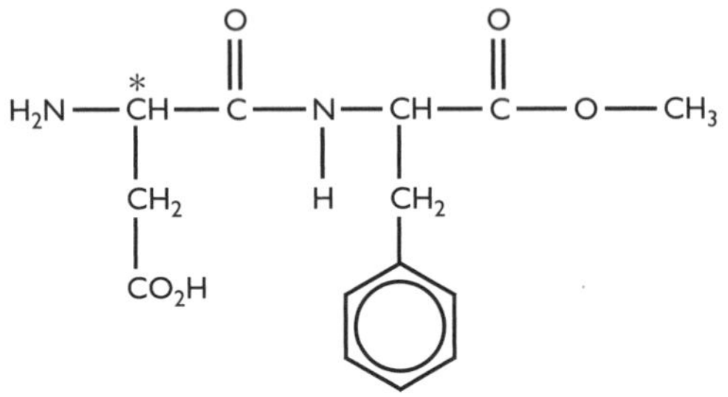

H$_2$N — CH — C(=O) — N — CH — C(=O) — O — CH$_3$

with CH$_2$—CO$_2$H branch (marked *) and H, CH$_2$—phenyl branch

> Candidate A gains both marks, but Candidate B has ignored the instructions in the question and identified only one of the chiral carbons.

Candidate A

(b) (ii) The carbon is bonded to four different atoms or groups. There will be four (2^2) stereoisomers.

Candidate B

(b) (ii) Carbon is bonded to four atoms or groups, and there will be four isomers.

> Candidate A gains both marks, but Candidate B carelessly loses the mark. Almost all carbon atoms are bonded to four atoms or groups. The key point missed by Candidate B is that the four atoms/groups are all *different*.

Candidate A

(c) (i) —

Candidate B

(c) (i)

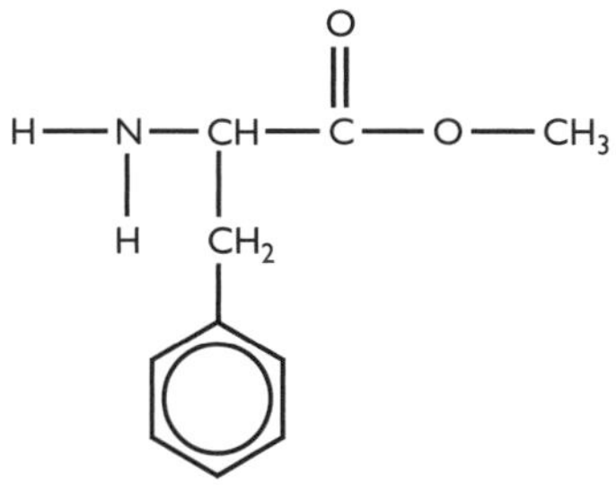

$$H-N-CH-C(=O)-O-CH_3$$

with H and CH_2 below, and a benzene ring below CH_2

🖉 Candidate A has made no attempt at this part, whereas Candidate B has used the information given in the question and deduced correctly the structure of the compound.

🖉 **Candidate B's response to the final part demonstrates considerable understanding, and yet his/her overall score is 7 out of 14 marks, while Candidate A gains 10. Look back at Candidate B's responses and identify where careless errors were made.**

Azo dyes

Time allocation: 8–9 minutes

(a) Nitrobenzene can be converted into benzenediazonium chloride ($C_6H_5N_2Cl$). State the reagents and conditions and write an equation for each step. Show the structure of the organic products. (6 marks)

(b) Benzenediazonium chloride reacts with a chlorinated phenol to form an azo dye with a relative molecular mass of 267 and the following composition by mass: **C, 53.9%; H, 3.0%; N, 10.5%; Cl, 26.6%; O, 6.0%.** Use this information to deduce the structure of the azo dye. (4 marks)

Total: 10 marks

■ ■ ■

Candidates' answers to Question 8

Candidate A

(a) Reagents: concentrated HCl and Sn

Conditions: refluxed

Equation: $C_6H_5NO_2 + 6[H] \rightarrow C_6H_5NH_2 + 2H_2O$

Reagents: sodium nitrite and hydrochloric acid

Conditions: excess HCl(aq) and the temperature must be below 10 °C

Equation: $C_6H_5NH_2 + HNO_2 + HCl \rightarrow C_6H_5N_2Cl + 2H_2O$

Candidate B

(a)

Both candidates clearly know their chemistry, but both have lost marks because of poor examination technique. Before starting such questions, you should read the question carefully and work out where the marks are likely to be allocated.

There are two steps for 6 marks, hence 3 marks each. You are asked for reagents and conditions ✓, an equation ✓, and the structure of the organic products ✓.

Candidate A has displayed good examination technique, but has scored only 2 marks for each step. Candidate A hasn't shown the structure of phenylamine or the diazonium chloride and therefore scores 4 out of 6 marks. Candidate B scores just 3 marks. In the first step, the equation is correct, as is the structure of phenylamine, so 2 marks are awarded. The second equation is also correct. However, the charge on the nitrogen in the diazonium compound is on the wrong nitrogen atom and there is no reference to the temperature required; hence just 1 mark is awarded.

The correct structure for the diazonium chloride is:

Candidate A

(b)

	C	H	N	Cl	O
Moles	53.9/12 = 4.5	3.0/1 = 3.0	10.5/14 = 0.75	26.6/35.5 = 0.75	6.0/16 = 0.375
Ratio (divide by smallest)	12	8	2	2	1
Mass of element in compound	144	8	28	71	16

Mass of compound $= 144 + 8 + 28 + 71 + 16 = 267$

Therefore, the empirical and molecular formulae are both $C_{12}H_8N_2Cl_2O$.

Candidate B

(b) $C_{12}H_8N_2Cl_2O$

Both candidates score 3 marks. Candidate B successfully deduces the structure of the azo dye (the most difficult part of the question), while Candidate A makes no attempt to do so. Candidate A demonstrates good examination technique, showing all working. Candidate B calculates the empirical formula, but does not use the relative molecular mass (267) to show that the molecular formula is the same as the empirical formula.

Candidate A has scored 7 out of 10 marks, while Candidate B scores only 6. This might not seem very different but, if this were repeated throughout the paper, then Candidate A would be awarded a grade B, while Candidate B would gain a grade C. Once again, poor examination technique has cost marks.

Question 9

Infrared spectroscopy and mass spectrometry

Time allocation: 11–12 minutes

(a) (i) Compounds **A** and **B** are structural isomers.

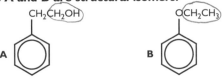

A   **B**

The mass spectrum of one of the compounds is shown below.
Explain how the fragmentation pattern allows you to deduce that
it is *not* compound **B**.   (2 marks)

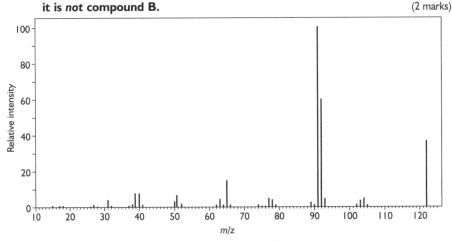

(ii) Write an equation to show the formation of the molecular ion for
compound **A**.   (1 mark)

(b) One of the compounds **A** or **B** gives the infrared spectrum below.

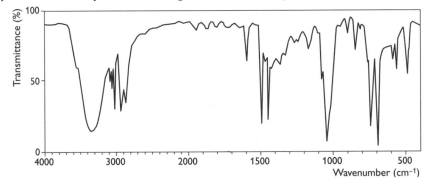

Using the data sheet, identify which of the two compounds, **A** or **B**, has this
spectrum. Explain your reasoning carefully.   (2 marks)

(c) **One of the compounds A or B gives the ¹H NMR spectrum below.**

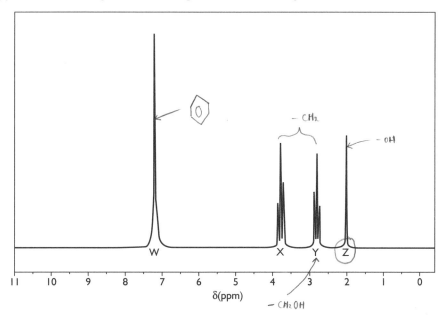

When a second spectrum was run with D_2O added, the peak Z, at δ 2.0, disappeared. Using the data sheet, suggest the identity of the protons responsible for the groups of peaks W, X, Y and Z. For each group of peaks, explain your reasoning carefully. Use *all* of the information given.

(1 mark is available for the quality of written communication.) (9 marks)

Total: 14 marks

■ ■ ■

Candidates' answers to Question 9

Candidate A

(a) (i) The side-chains will break off and compound A will form ions due to CH_2OH at $m/z = 31$, whereas compound B will form an ion at 29 due to CH_2CH_3, which is not present in the spectrum.

Candidate B

(a) (i) B would form an ion at 29, and it is not there.

> ✏ Candidate A gives a textbook answer and gains both marks. Candidate B clearly understands fragmentation, but has not explained what would be responsible for a peak at 29, and so scores only 1 out of 2 marks.

9

question

Candidate A

(a) (ii)

$$\text{C}_6\text{H}_5\text{—CH}_2\text{CH}_2\text{OH} \longrightarrow \text{C}_6\text{H}_5\text{—CH}_2\text{CH}_2\text{O}^- + \text{H}^+$$

Candidate B

(a) (ii)

$$\text{C}_6\text{H}_5\text{—CH}_2\text{CH}_2\text{OH} + \text{e} \longrightarrow \left(\text{C}_6\text{H}_5\text{—CH}_2\text{CH}_2\text{OH}\right)^+ + 2\text{e}$$

🖉 Candidate B gains the mark, but Candidate A has misunderstood the process of forming the molecular ion, and scores nothing.

Candidate A

(b) The peak between 3230 and 3550 cm^{-1} is due to the hydrogen bonding in alcohols and therefore the spectrum is for compound A.

Candidate B

(b) It's the alcohol.

🖉 Candidate A gives the perfect answer, making use of the instructions in the question and quoting directly from the data sheet. Candidate B is awarded 1 mark for correctly identifying the spectrum as that of an alcohol. However, the other mark is lost for not explaining *why* it is an alcohol and for not stating which of compounds A or B is the alcohol.

Candidate A

(c) When the spectrum is re-run using D$_2$O, peak Z disappears, showing that peak Z is due to a labile proton that is found in groups such as the —OH group. Peaks X and Y are both split into triplets, showing that the adjacent carbons must be attached to two hydrogens ($n + 1$) rule. Peak W is due to the five hydrogens on the benzene ring.

Candidate B

(c) It has to be compound A, because the alcohol OH peak disappears when run in D$_2$O. Compound B doesn't have an OH.

🖉 The marking points are: peak W — benzene ring ✓; chemical shift = 7.1–7.7 ppm or relative number of hydrogens = 5 ✓. Peaks X and Y — both are CH$_2$ ✓; both are split into triplets ✓; adjacent carbon atoms are attached to two hydrogen atoms✓. Peak X — CH$_2$ attached to the benzene ring 2.3–2.7 ✓; Peak Y — CH$_2$ attached to O 3.3–4.3 ppm ✓; Peak Z — OH ✓. Reasons: when re-run in D$_2$O it disappears ✓.

🖉 The quality of written communication mark is awarded for the correct use of specific chemical terms. Any two from 'labile', 'chemical shift' or 'splitting' gain the mark. ✓

Candidate A has tried to follow the guidelines given in the question and has scored a total of 6 marks. Try marking Candidate A's answer and see if you can identify which 3 marks have been lost. Candidate B clearly understands NMR and can use it to identify the correct compound, but ignores most of the question. This poses a real dilemma for the examiner. The marker has to stick to the agreed mark scheme, and it is just about possible to award Candidate B 2 marks. However, the candidate neither follows the instructions nor identifies the groups responsible for any of the four peaks.

Candidate A scores 10 out of 14 marks, while B scores 5 out of 14 marks. Again, Candidate A demonstrates better examination technique. This is most apparent in the final part of the question, which is open-ended and allows for a free response. Candidates have to use the information and the number of marks allocated to the question, and they must plan how best to gain those marks. Usually, 1 mark is given for making one correct point, so a score of 9 marks requires nine separate points.

NMR spectroscopy

Time allocation: 3–4 minutes

(X) A sample of torn clothing was found at the scene of a crime. The clothing had a yellow stain, which was thought to be either:

Compound **A**   or   Compound **B**

Explain how the forensic scientist could determine which, if either, of compound A or compound B was responsible for the yellow stain. **(5 marks)**

(b) Suggest a chemical test that could be used to distinguish between compounds A and B. State the reagent and conditions and explain what you would expect to see. Identify the organic product. **(3 marks)**

Total: 8 marks

■ ■ ■

Candidates' answers to Question 10

Candidate A

(a) The forensic scientist should obtain a mass spectrum of each compound and find the molar mass. Compound A would have a higher mass than compound B. The carbon-13 NMR of compound A would have more peaks than the ^{13}C NMR of compound B. The infrared spectrum of compound A would have a broad peak around 3200–3500 due to the alcohol, —OH, and the other one wouldn't.

Candidate B

(a) The ^{1}H NMRs of both compounds would have some peaks in common and it would be difficult to distinguish between them. If D_2O was used, then two peaks would disappear for compound A and only one peak for compound B would disappear. Compound A would react with a carboxylic acid to make an ester, but compound B wouldn't.

> ⚡ This is a difficult question, as it is open-ended and there are many ways to respond. However, both candidates miss the point of the question and explain how compounds A and B differ rather than how either could be identified unambiguously. Both candidates score marks because both have made some correct statements. Candidate A has recognised that mass spectra have different molecular-ion peaks and that the ^{13}C NMRs would have a different number of peaks. Therefore, Candidate A would probably score 2 out of 5 marks. However,

Candidate B would probably gain only 1 mark, because he/she has recognised that ^{1}H NMR could be used to distinguish between them.

Unambiguous identification can be achieved only by comparison of the spectra with those from a database. The fingerprint region of the infrared spectrum is helpful, but may not be conclusive. If the infrared is used in conjunction with GC–MS, then, by using the fragmentation pattern of the mass spectrum, it should lead to an unambiguous identification.

Candidate A
(b) Compound A contains an alcohol, so it will react with sodium and give off bubbles of hydrogen. The organic product would be:

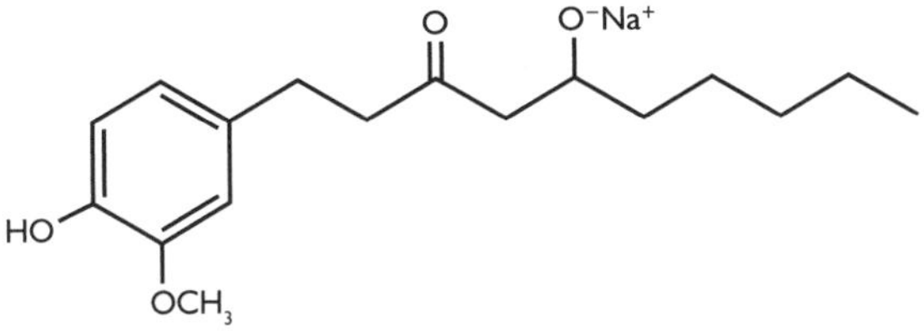

Candidate B
(b) Compound A can be oxidised by $H^+/Cr_2O_7^-$, and it will turn from orange to green. The product is:

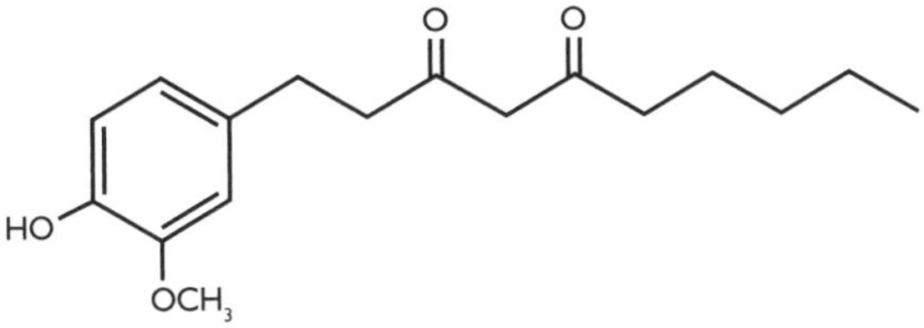

Both candidates would probably score 2 out of 3 marks. Candidate A loses a mark for suggesting that sodium could be used. Sodium does react with the alcohol group, but it also reacts with a phenol. Compounds A and B are both phenols. Candidate A would score a mark for the observation and might gain a mark for the product, although the correct product should be:

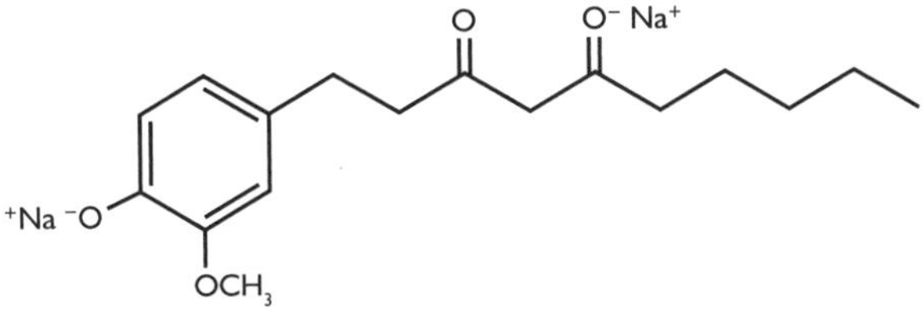

Candidate B very nearly scores all 3 marks, but carelessly loses a mark by writing the dichromate ion as $Cr_2O_7^-$, when it should have been $Cr_2O_7^{2-}$.

This was not a good question for either candidate, with Candidate A scoring 4 (possibly 3) out of 8 marks. Candidate B was awarded only 3 marks.

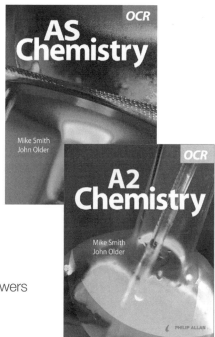